LATIN AMERICAN MODERN ARCHITECTURES

Latin American Modern Architectures: Ambiguous Territories has thirteen new essays from a range of distinguished architectural historians to help you understand the region's rich and varied architecture. It will also introduce you to major projects that have not been written about in English. A foreword by historian Kenneth Frampton sets the stage for essays on well-known architects, such as Lucio Costa and Félix Candela, which will show you unfamiliar aspects of their work, and for essays on the work of little-known figures, such as Uruguayan architect Carlos Gómez Gavazzo and Peruvian architect and politician Fernando Belaúnde Terry. Covering urban and territorial histories from the nineteenth and twentieth centuries, along with detailed building analyses, this book is your best source for historical and critical essays on a sampling of Latin America's diverse architecture, providing much-needed information on key case studies.

Contributors include Noemí Adagio, Pedro Ignacio Alonso, Luis Castañeda, Viviana d'Auria, George F. Flaherty, María González Pendás, Cristina López Uribe, Hugo Mondragón López, Jorge Nudelman Blejwas, Hugo Palmarola Sagredo, Gaia Piccarolo, Claudia Shmidt, Daniel Talesnik, and Paulo Tavares.

Patricio del Real holds a PhD in Architecture History and Theory from Columbia University.

Helen Gyger holds a PhD in Architecture History and Theory from Columbia University.

LATIN AMERICAN MODERN ARCHITECTURES

Ambiguous Territories

Edited by Patricio del Real and Helen Gyger

Routledge
Taylor & Francis Group

NEW YORK AND LONDON

First published 2013
by Routledge
711 Third Avenue, New York, NY 10017

Simultaneously published in the UK
by Routledge
2 Park Square, Milton Park, Abingdon, Oxon OX14 4RN

Routledge is an imprint of the Taylor & Francis Group, an informa business

Library of Congress Cataloging in Publication Data
Latin American modern architectures: ambiguous territories/
[edited by] Patricio del Real and Helen Gyger.
    pages cm
    Includes index.
    1. Architecture—Latin America—History—20th century.
    I. Del Real, Patricio, 1966– editor of compilation.
    II. Gyger, Helen, editor of compilation.
    NA702.5.L39 2012
    720.98'0904—dc23
    2012007233

ISBN: 978-0-415-89345-9 (hbk)
ISBN: 978-0-415-89346-6 (pbk)
ISBN: 978-0-203-10128-5 (ebk)

Acquisition editor: Wendy Fuller
Editorial assistant: Laura Williamson
Production editor: Ben Woolhead

Typeset in Corbel
by Florence Production

Printed and bound in the United States of America by
Sheridan Books, Inc. (a Sheridan Group Company).

CONTENTS

FOREWORD

Kenneth Frampton

This remarkable collection of essays emerging out of the symposium, *Ambiguous Territories: Articulating New Geographies in Latin American Modern Architecture and Urbanism*, held at the Graduate School of Architecture, Planning, and Preservation at Columbia University on March 27 and 28, 2009, may be taken as an indicator that Latin America is a continent whose time has indisputably arrived after half a century of provincial amnesia and neglect. One has to go back to Philip Goodwin and G. E. Kidder Smith's *Brazil Builds* (1943), Henry-Russell Hitchcock's *Latin American Architecture since 1945* (1955), or to Andre Bloc's editorship of *L'Architecture d'aujourd'hui* in the 1950s to find comparable international interest in Latin American architectural culture and even then, as the editors of this anthology imply, the documentation and analysis in this period invariably tended to assume a simplistic laudatory tone or, occasionally, a militant, nationalistic perspective put together by an earlier generation of international critics and scholars. Here instead we have a set of newly emergent scholars from Latin America and beyond, who have not only mapped these ambiguous territories, but also cultivated a different critical optic capable of passing with revelatory effect from tightly focused local cultural developments to ideological and political manifestations having a wider transnational scope.

This anthology is subdivided for editorial convenience into three different categories; focusing first, on unfamiliar cultural trajectories pursued by exceptional individuals; second, on the various forms assumed by the overarching thrust of modernization throughout the continent; and third, on the various overlaps and interactions between quite different national territories and policies.

Irrespective of this taxonomy, the complex labyrinthine process of modernizing culture-politics permeates every one of the chapters, as we may judge if we randomly compare Gaia Piccarolo's study of Lucio Costa's inadvertent reversal of the time-honored hierarchy between center and periphery to Daniel Talesnik's account of the realization of the highly symbolic UNCTAD III convention center in Santiago during Salvador Allende's tragically eclipsed tenure as president of Chile, or to María González Pendás's analysis of Félix Candela's hitherto unknown "geopolitical imagination."

In one way or another, most of this material treats with the modernization of the Latin American continent and with the tragic political and ecological consequences that this has often brought in its wake. This is never more evident than in Paulo Tavares's chapter on the development of the Amazon basin from the mid-1950s onward; the state-sponsored exploitation of the rainforest entailed among other egregious operations the surreptitious involvement of the CIA in the 1964 military seizure of power in Brazil and the ensuing "fish-spine" clearing of the jungle which led among other things to the so-called "humanistic" clothing and sequestering of the indigenous population and the engagement of such distinguished modern architects as Oswaldo Bratke in the contradiction of planning a "utopian" steel company town in the midst of the Amazonian delta.

Of all of the cross-fertilizations to have visited Latin America, none perhaps is more surprising and inspiring than the persistent affinity oscillating across the Atlantic between Iberia and its former colonies divided linguistically between Mexico, in the first instance, and Brazil in the second. With the first, while writing about the self-made engineer/constructor Candela, González Pendás recalls the refugee coterie of leftist intellectuals exiled in Mexico after the denouement of the Spanish Civil War, while with respect to the second, Piccarolo reveals how a right-wing transatlantic vision of progress involved the mutual search for an aboriginal Portuguese vernacular to serve as a legitimizing nationalistic trope to be patronized to an equal degree by both Getulio Vargas's Estado Novo of 1937 and Salazar's much more contradictory conservative modernizing policies being pursued in Portugal at around the same time. In the case of Mexico, we have the compensatory paradox of Félix Candela's vision of a progressive transatlantic Hispanic federation in which an all but mythical, modern, techno-economic model was envisaged as being exported back to Spain, whereas in the case of Brazil, we have Lucio Costa's concrete research into the eighteenth-century colonial vernacular of Minas Gerais being transferred through a kind of cultural osmosis in which a similar line of research was carried out by the Portuguese Syndicate of Architects, in the name of the state, and published in 1961 under the title *Arquitectura popular em Portugal*.

A more abstruse attempt at an authentic modern national identity based on tradition came about in the 1920s, as Cristina López Uribe has shown, with the Mexican adaptation of an ersatz Spanish Colonial manner, which can be dated back to Bertram Goodhue's contribution to the Panama-Californian Exposition of 1915. This style came to be widely adopted in Mexico and elsewhere in Latin America throughout the roaring twenties, a genre whose popularity was greatly enhanced by the exotic hacienda settings of Hollywood that were evident both on the screen and off. Since the cinema was regarded at the time as the quintessence of modernity, the Mexican *Colonial Californiano* proffered itself as an alternative modernism, one which embodied a more evocative and romantic iconography than the flat-roofed asperities of the International Style. And we may say, as the author implies, that Luis Barragán's inimitable mature manner came about as a subtle and sensitive synthesis between these two otherwise totally antithetical modes.

Much later with the subsequent petrified stalemate of the Cold War, a more overtly productive and materialist worldview had a certain impact on Latin America through the Soviet provision of their homegrown prefabricated housing systems to two successively emergent socialist states: first to Cuba, soon after the revolution and the subsequent hurricane which devastated the eastern part of the island in 1963 and then, a decade later in 1972, with the building of a prefabricated panel factory in Chile during the brief, democratically elected Socialist regime of Salvador Allende. As Pedro Ignacio Alonso and Hugo Palmarola Sagredo inform us, this factory produced some 153 housing blocks before Pinochet's US-backed coup d'etat put an abrupt and violent end to this promising enterprise.

After the Second World War, a succession of master architects and/or politicians— with the two being occasionally combined in the same person—had a decisive impact on the development of Latin America. As far as architect-políticos are concerned, two of the most prominent in the postwar period were Pedro Ramírez Vázquez of Mexico, and Fernando Belaúnde Terry, the latter being twice elected president of Peru—first in 1963 and then in 1980. As Luis Castañeda informs us, like Vargas before them, both men understood architecture and regional planning as enacting essential roles not only with regard to the formation of national identity but also with respect to generating a significant modernizing economic agenda. It is possible to characterize the influence of certain well-known modern architects in similar terms, particularly when it comes to their impact on the development of a number of countries within the overall framework of the continent; above all, perhaps as these studies demonstrate, the seminal roles played by such figures as Mario Pani in Mexico and Carlos Raúl Villanueva in Venezuela, roles that are prefigured, as Jorge Nudelman reveals, in the architectural and political culture of 1930s Uruguay. We now know, with Viviana d'Auria, by being wise after the event, that these progressive interventions, both theoretical and practical, had long-term repercussions that were, in the end, not so desirable, as is evident in the case of the eighty-five high-rise *superbloques* built around Caracas between 1954 and 1958 to the designs of Villanueva. We should tell ourselves once again, as this remarkable anthology repeatedly demonstrates, that as far as Latin America is concerned, it was never possible to separate architecture from politics or from the rigorous modernizing drive which still makes up the intrinsic substance of its spirit.

CONTRIBUTORS

Noemí Adagio has a degree in Architecture from the Universidad Nacional de Rosario (1982), and a Master's degree from the École d'Architecture Paris-Villemin (1987). She is a full-time researcher in the Facultad de Arquitectura, Planeamiento y Diseño at the Universidad Nacional de Rosario, where she teaches architectural history. In addition she heads the contemporary architecture theory seminar in the Master's program at the Facultad de Arquitectura y Urbanismo at the Universidad Nacional del Litoral in Santa Fe. Her area of specialization is the history of twentieth-century architecture, and her most recent publications focus on notable figures and issues in architectural modernism.

Pedro Ignacio Alonso holds a Master's in Architecture from the Pontificia Universidad Católica de Chile (PUCCh) in Santiago and completed his PhD at the Architectural Association (AA) in London. He has taught at the AA since 2005, most recently as Visiting Tutor in the History and Critical Thinking Master's program, and teaches design and architectural theory at PUCCh in Santiago. Together with Hugo Palmarola, he received a RIBA Research Trust Award in 2008 for research on Soviet systems of prefabrication and their reception in Cuba and Chile during the Cold War. In 2010, he received a grant from the Getty Research Institute in Los Angeles, and in 2011 a Fellowship as Visiting Scholar at the Canadian Centre for Architecture.

Luis Castañeda is an Assistant Professor of Art History at Syracuse University, New York. He received his PhD in Twentieth-century Art and Architecture from the Institute of Fine Arts, New York University, in 2011. His dissertation discusses the design campaign for the 1968 Mexico City Olympics. He is currently working on a book project that will explore the intersections between the production of official culture and processes of industrialization and urban modernization in mid-twentieth-century Mexico and Latin America more broadly.

Viviana d'Auria trained as an architect in Rome, and is currently a PhD candidate at the Department of Architecture, Urbanism, and Planning (ASRO) of the Katholieke Universiteit Leuven, Belgium, where she also earned a Master's degree in Architecture of Human Settlements. Her research investigates the epistemological contribution of developmentalism and technical assistance projects to urbanism, with a particular focus on Venezuela and (post)colonial Ghana. She is a co-editor of *Human Settlements: Formulations and (re)Calibrations* (SUN Academia, 2009) where "Third World" cities are explored as sites for experimental practices and knowledge exchange.

Patricio del Real trained as an architect at Harvard University's Graduate School of Design and received a PhD in Architecture History and Theory from Columbia University, New York. His writings on modernity in Latin America have appeared in several journals; his current research focuses on the construction of a Latin American imaginary through modern architecture during the early years of the Cold War. He has taught architecture since 1991 in the Americas and Europe.

George F. Flaherty trained as an art historian, and specializes in Latin American and US Latino visual and spatial cultures since 1945. He is working on a book manuscript that explores the spatial dimensions of the 1968 Tlatelolco massacre in Mexico City and its mediation. He has held fellowships from the Center for Advanced Study in the Visual Arts, the Social Science Research Council, the Society of Architectural Historians, and the US-Mexico Fulbright Commission. A graduate of Swarthmore College and the University of California, Santa Barbara, he joined the Art History Faculty at the University of Texas at Austin in fall 2011.

Kenneth Frampton trained as an architect at the Architectural Association in London. After practicing for a number of years in the United Kingdom and in Israel, he served as the editor of the British magazine *Architectural Design*. He has taught at a number of leading institutions including the Royal College of Art, the ETH Zurich, EPFL Lausanne, the Accademia di Architettura in Mendrisio, and the Berlage Institute in the Netherlands. He is currently the Ware Professor of Architecture at the Graduate School of Architecture, Planning, and Preservation, Columbia University, New York. He is the author of *Modern Architecture and the Critical Present* (1980), *Studies in Tectonic Culture* (1995), *American Masterworks* (1995), *Le Corbusier* (2001), *Labour, Work & Architecture* (2005), and an updated fourth edition of *Modern Architecture: A Critical History* (2007).

María González Pendás is a PhD candidate in Architecture History and Theory at the Graduate School of Architecture, Planning, and Preservation, Columbia University. She graduated as an architect from the Escuela Técnica Superior de Arquitectura de Madrid, and worked in several architectural offices in Madrid and Chicago before being granted a Fulbright fellowship to undertake graduate studies in the United States. Her writing focuses on the question of modern architectural realisms, and on the dialectics between processes of modernization of architecture, politics, and society, especially during the second half of the twentieth century. Her publications include "Paris-Nord: Shadrach Woods's Imaginary Global City" (co-authored with Patricio del Real, 2010) and "Apatridas Arquitecturas: Félix Candela, José Luis Sert, and the Return of the Modern to Postwar Spain" (forthcoming).

Helen Gyger holds degrees in visual arts from Sydney College of the Arts, a Master's from the New School for Social Research, New York, and a PhD in Architecture History and Theory from Columbia University. Her research has been supported by grants from the Graham Foundation for Advanced Studies in the Fine Arts, the Society of Architectural Historians, and the Paul Mellon Centre for Studies in British Art. She is currently working on a book project developing out of her dissertation, *The Informal as a Project: Self-Help Housing in Peru, 1954–1986*.

Cristina López Uribe graduated as an architect from the Universidad Nacional Autónoma de México (UNAM) in 2001. She holds a Master's degree in the Theory and History of Architecture from the Universitat Politécnica de Catalunya, in Barcelona (2002). In 2008, she was the recipient of a grant from the Consejo Nacional Para la Ciencia y Tecnología (CONACYT) to complete her doctoral degree in Theory and History of Architecture at the Universitat Politécnica de Catalunya. Her PhD research focuses on the early works of Mexican architect Juan O'Gorman, and on the arrival and adaptation of the ideals of the Modern Movement in Mexico City in the first decades of the twentieth century.

Hugo Mondragón López is currently the head of the Master's of Architecture program at Pontificia Universidad Católica de Chile (PUCCh) and director of *CA: Ciudad y Arquitectura*, the official magazine of the Chilean Architects' Association, and is an active member of Docomomo Chile. He graduated as an architect from Universidad Piloto de Colombia (1990), holds a Master's in Architecture, PUCCh (2002), and a PhD in Architecture and Urban Studies, PUCCh (2010). Mondragón's research interests are focused on the interrelations between modernization and modernity, primarily as shown in architectural journals.

Jorge Nudelman Blejwas trained as an architect at the Escuela Técnica Superior de Arquitectura de Barcelona, and holds a doctorate from the Escuela Técnica Superior de Arquitectura de Madrid. He is currently a Professor of Architectural Theory at the Instituto de Historia de la Arquitectura, Facultad de Arquitectura, Universidad de la República, Uruguay (UDELAR), and also teaches architectural design at UDELAR and the Universidad ORT Uruguay in Montevideo.

Hugo Palmarola Sagredo studied design at the Pontificia Universidad Católica de Chile (PUCCh), where he taught subsequently between 2004 and 2007, and holds a Master's degree in Histories and Theories of Industrial Design from the Universidad Nacional Autónoma de México (UNAM). Between 2007 and 2009 he was a guest of the International Scholars Program of the Society for the History of Technology (SHOT). He has published widely—in Latin America, the United States, Europe, and Asia—on the history of design in Chile and is currently conducting PhD research in Latin American Studies at UNAM on the processes of technological assimilation in the domestic realm in Latin America during the first half of the twentieth century.

Gaia Piccarolo is an architect and received her PhD in the History of Architecture and Urbanism from the Politecnico di Torino in 2010. She is Adjunct Professor of Architectural History in the Contemporary Age at the Politecnico di Milano. She has contributed to the organization and design of several exhibitions and currently writes for *Il Giornale dell'Architettura* and *Paesaggio Urbano*. She is the author of various essays on the circulation of ideas and models within architectural culture at the end of the nineteenth century and the first half of the twentieth century, with special focus on Italy and the Luso-Brazilian world. Her research has been presented in Europe, the United States, Canada, and Brazil in the framework of international events and congresses. In 2007, she was co-editor with Giuliana Ricci of the book *Luigi Manini (1848–1936), architetto e scenografo, pittore e fotografo*.

Claudia Shmidt is an architect and holds a PhD in History and Theory of the Arts from the Universidad de Buenos Aires. She is currently director of the Master's program in the History and Culture of Architecture and the City at the Escuela de Arquitectura y Estudios Urbanos of the Universidad Torcuato Di Tella, Buenos Aires, and is also a member of the editorial committee of the journal *Block*. Her dissertation focused on Buenos Aires as a case study to examine the role of public architecture in the articulation of culture, politics, and the city in the trans-formations of new capitals of emerging nation-states in the second half of the nineteenth century. Her current research focuses on public architecture produced by the state and on the history of theories of architectural space from the end of the nineteenth century to the present.

Daniel Talesnik holds a Master's degree in Architecture from the Pontificia Universidad Católica de Chile (PUCCh), a Master's in Advanced Architectural Design from the Graduate School of Architecture, Planning, and Preservation, Columbia University (2008), and is currently a PhD candidate in Architecture

History and Theory at the GSAPP. His current work centers on politically driven European architectural projects of the first half of the twentieth century, and their repercussions outside of Europe—specifically in Latin America.

Paulo Tavares is an architect and urbanist who trained in Brazil, and received his Master's from the Centre for Research Architecture, Goldsmiths College, University of London, where he currently teaches in the Master's program. His work inhabits the interfaces between architecture, media, and ecology. He lives in São Paulo and London.

ACKNOWLEDGMENTS

This project developed out of a conference that we organized at Columbia University in March 2009, titled *Ambiguous Territories: Articulating New Geographies in Latin American Modern Architecture and Urbanism*, and we would like to begin by thanking the various organizations that sponsored that event: the Graduate School of Architecture, Planning, and Preservation at Columbia University, and Dean Mark Wigley; the Temple Hoyne Buell Center for the Study of American Architecture, and its director, Reinhold Martin; and the Graham Foundation for Advanced Studies in the Fine Arts.

We would also like to thank all those who participated in the conference as panel moderators—Kenneth Frampton, Barry Bergdoll, Daniel Barber, María González Pendás, and Jorge Otero-Pailos; as well as presenters—Eduardo Mendieta, Umberto Bonomo, Federico Deambrosis, Andrea Y. Flores Urushima, Renato Holmer Fiore, Guillermo Jajamovich, Salvador Lizárraga Sánchez, Johanna Lozoya, Ricardo Hernán Medrano, María Isabel Oliver, Ricardo Rocha, and André Tavares; and finally our colleagues in the PhD Program in Architecture History and Theory at Columbia University who helped to organize the conference, along with Diana Martinez and Anna Kenoff of the Buell Center.

INTRODUCTION

Ambiguous Territories

Patricio del Real and Helen Gyger

> Latin America. Two words, which in Europe have been and are exploited in all manners imaginable of ruthless ambition: Latin America . . .
>
> Latin America lends itself to speeches, verses, tales, film festivals with music, food, and beverages, and Sunday entertainment. In the name of Latin America plundering is on the rise.
>
> (César Vallejo, Favorables Paris, 1926)

In his 1972 review of Francisco Bullrich's *New Directions in Latin American Architecture* (1969), Reyner Banham commented on the inevitable collapse of any examination of architecture at a "Continental spread." Such surveys of the "State of the Art" covering vast regions as diverse as Latin America were prone, he argued, to be at best fragmented, and at worst superficial; ultimately, they represented a "Continental failure."[1] It is clear that many of the limitations of Bullrich's panoramic work were due to the format of Braziller's *New Directions in Architecture* series itself—and this was not a uniquely Latin American flaw, since Banham perceived a similarly failed continentalism in the volumes on *African Architecture* by Udo Kultermann and *American Architecture* by Robert Stern—but this does not entirely explain Banham's critique. If the responsibility did not lie with the author Bullrich—redeemed in Banham's eyes by his criticality and his

INT.01 Bullrich, *New Directions in Latin American Architecture*: Félix Candela, "Umbrellas," Coyoacán Market, Mexico; Ricardo Porro, Plastic Arts School, Havana, Cuba, 1962–1965

authority "as the virtual father of modern architectural history in Argentina"—then perhaps it was due to a historical impasse which made such editorial "round-ups" untenable; or perhaps, as César Vallejo pointed out almost fifty years before, it was the price to pay in the global consumption of branded goods.

Yet Banham was less concerned with architecture's inability to fulfill the unity announced in the term "Latin America"—to perform a singular, distinctive regional identity in contrast to the modernist universalism already articulated from the privileged position of Western Europe—than with the misbegotten enterprise of attempting to write such a history on its behalf. Indeed, the production of a "Latin American" modern architecture has rarely (if ever) been the ambition of the region's architects, but rather a framework that from time to time has been projected onto architecture by historiography—one of a range of cultural identities defined by variable locational constructions: whether nation-state boundaries (Nationalism), the specificities of local customs and topographies (Regionalism), shared language and historical experience (Hispano-Americanism), physical proximity (Pan-Americanism), or cultural and intellectual connections which are seen to transcend geography (Cosmopolitanism). In architecture, these shifting territories of cultural identification have been reflected in competing formal languages—Beaux-Arts, neocolonial, Art Deco, Functionalism—deployed in debates over the relative values of tradition or newness, of indigeneity or imported influence, provoked by the intensifying pressures of technological, economic, and social changes brought about by modernization. Behind each of these stylistic expressions there was a project of modernity—articulating plural ways of being modern. Rather than a simple linear progression from one style to the next, the decades around the turn of the twentieth century were characterized by oscillations between these various positions; these architectural modernities were synchronous rather than sequential and discrete.

Is there anything to this term "Latin America"? Dismissed as a hall of mirrors, an essentializing cultural concept, a failed ideal of political unity—has its symbolic power and elusive meaning been exhausted? While it is clear what we gain by abandoning this category—ridding ourselves of oversimplifications and over-generalizations, the tendency to search for superficial resemblances and convergences—we believe that there is also something to be lost by a refusal to

INT.02 Bullrich, *Arquitectura latinoamericana*, *1930–1970*: "Past and Present" with view of Monte Alban, Mexico

engage with the "continental" frame. According to Banham's own assessment, Bullrich's endeavor to draw supra-national comparisons—between Eladio Dieste in Uruguay, Félix Candela in Mexico, Victorio Garatti and Ricardo Porro in Cuba, for example—was both "enterprising" in its ambition and "salutary" in its results. In such cases, the broader continental frame has the potential to stretch our understandings of history beyond the limitations of the national into more ambiguous territories, illuminating moments of intra-regional dialogue around specific challenges and solutions, or revealing shared particularities in the local inflections of an "International Style."

Rather than inevitably creating an essentializing category that homogenizes the entire region, then, we argue for the possibility of "Latin America" as a pathway into exploring a network of transnational relationships and transcontinental connections. "Latin America" in the plural, not the singular—just as its modernities in architecture have operated in the plural, not the singular. While Banham regretted that Bullrich's survey was ultimately "a bit fragmented"—despite (or because of) his efforts to narrate a panoramic history—from our perspective four decades later, the prospect of offering a "fragmented" survey of Latin American modern architectures is both inevitable, and liberating. Assembled as if in a collage, these histories are not subsumed under pre-existing narratives, but invite multiple points of entry into the architecture of the region.

We begin with a brief discussion of the term "Latin America," before turning to the questions of how Latin American modernity has been framed and given meaning in the region's architectural culture, and how the Latin American modern has been perceived from the outside. In conclusion, we explore how this book aims to expand, reframe, and offer new perspectives on the ambiguous territories of Latin American modern architectures.

América

If there is such a thing as Latin American architecture, there must be a place called Latin America. The term "Latin" America first appeared in print in 1836 in Michel Chevalier's introduction to his *Lettres de l'Amérique du Nord*. Drawing on an older reading of Europe as divided between two great races—the Latin and the Germanic—Chevalier described the difference between the United States and the countries to its south in identical terms: "South America is, like southern Europe, Catholic and Latin. North America belongs to a Protestant and Anglo-Saxon population."[2] Subsequently, France's deployment of this binary opposition was undeniably politically motivated: having positioned itself as the political and cultural leader of "Latin Europe," it then sought to extend its influence into the Americas by promoting this apparent cultural affinity with the southern nations. Yet the popularization of the term was due to the fact that for many within the region the descriptor had a strong appeal, suggesting a family connection to the cultural and intellectual achievements of a modern European power which had the advantage of not being the former colonial master—the experience of Haiti aside.[3] The notion of an *Amérique latine* was consolidated with the French "expedition" to Mexico, which in 1864 had installed Maximilian of Habsburg as head of a new Mexican Empire.[4] Fifty years earlier, a precedent of sorts had been set with the 1816 French artistic mission to the Brazilian Empire headed by architect Grandjean de Montigny, which resulted in the creation of the Imperial School of Arts in Rio de Janeiro and a new urban landscape of neoclassical monuments to enhance the capital's efforts to reinforce its rule over what was still a country of scattered coastal cities.[5] In Brazil—as in much of Latin America— this French cultural dominance continued throughout the nineteenth century and into the twentieth with the influence of the École des Beaux-Arts on architectural design, education, and urbanism; this in turn formed an important backdrop to Le Corbusier's much-celebrated visit to the region in 1929.

Establishing a collective identity for "Latin" America was seen to make it stronger in the face of an increasingly powerful northern neighbor, as the United States sought to expand and assert its influence in the region beginning with the Monroe Doctrine in 1823. In the wake of the Spanish-American War (1898), Uruguayan writer José Enrique Rodó played out this self-consciously oppositional identity in his allegory *Ariel* (1900), where the figure of a spiritual Latin America—Ariel—is threatened by the crudely materialistic Caliban, exemplified by the United States; this trope still recurs in discussions over the nature of a Latin American identity.[6] From the viewpoint of the United States, the Spanish-American War marked the development of the study of "Latin America" as an academic discipline, providing a means to establish a body of institutional knowledge on its new protectorates and possessions.[7] The architectural expression of this tendency can be seen in Sylvester Baxter's introduction to his ground-breaking historical study *Spanish-Colonial Architecture in Mexico* (1901): with the United States now "taking over" diverse former Spanish territories (a process initiated in 1846 with the US-Mexico War), Baxter observed that this made "the study of Spanish and Spanish-Colonial architecture of particular interest to our architects, whose services will naturally be increasingly demanded in connection with the development that will necessarily attend the changing conditions in those lands."[8] The United States would thus adopt the costume of this "Latin" identity in order to fulfill its role as (neo)colonial administrator. More broadly, the late arrival of the United States to the game of imperialism reshaped the American continent, most dramatically through the construction—both physical and political—of the Panama Canal; US intervention was sustained by Theodore Roosevelt's 1904 Corollary to the Monroe Doctrine and the logic of a "sphere of influence" which has had a long and active history.

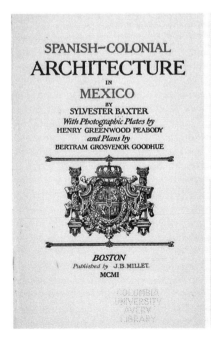

INT.03 Baxter, *Spanish-Colonial Architecture in Mexico*: Village Church at Ixtacalco; Puente de la Laja, Celaya. Tresguerras, architect

Does this antagonism between north and south, between an Anglo-Saxon and a Latin America still hold today? If "Latin" America was defined from the outset in opposition to the north, does the term remain a dominated signifier—"a word image defined by power structures that frame the relationship between the United States and the rest of the American continent"—as Chilean artist Alfredo Jaar proposed more than twenty years ago in his installation "A Logo for America" in Times Square, New York?[9] It could be argued that this antagonism has reduced cultural relations to a bipolar understanding, predetermining and conditioning much of the intellectual debate undergirding architectural scholarship; that it remains the core impediment to a pluralistic and dynamic definition of Latin America.

Ambiguous Modernities

The front and back covers of a recent collection of essays, *Latin American Architecture, 1929–1960* (2004),[10] display what have become two iconic images of the region's modernist architecture: the interior of Carlos Raúl Villanueva's Aula Magna at the Universidad Central de Venezuela (1950–1953) and Oscar Niemeyer's Palace of the Dawn, Brasília (1956–1958). The two form the centerpieces of major projects from the mature careers of Latin America's foremost architect-auteurs— state-sponsored projects with a strong developmentalist agenda, respectively the main lecture hall of the new national university and the presidential palace of the new national capital, embodying the humanist values of education and democracy.

But how exactly do these images work as visual metonyms for the region? Beginning with two key texts in the historiography of Latin American modern architecture, we examine its characteristics have been defined.

In *Latin American Architecture since 1945*, the catalogue accompanying his influential 1955 exhibition at the Museum of Modern Art (MoMA) in New York, Henry-Russell Hitchcock described an essentially international architectural language with some regional inflections, and—typical of a mid-century perspective —viewed "modern" and "modernism" as virtually synonymous. He noted the lingering effects of "official French taste," and that the region had not "produced important autochthonous developments" during the formative period of modernism. Nonetheless, "at the expense of oversimplification" it was possible to identify some common traits, primarily the result of climatic conditions and the qualities of local materials.[11] For example, since the region lacked adequate steel production and good structural timbers, and its natural stone and manu-factured bricks tended to be of inferior quality, there was an extensive use of stucco (often painted) to disguise these flaws; in this context, concrete was quite naturally enthusiastically embraced as a building material. Adaptations to climate included the *brise-soleil*, and the use of colored stucco or tiles to avoid the glare of pure white surfaces. One additional formal trait was ascribed to cultural rather than natural conditions: "Curved skylines such as segmental and paraboloid forms produce are far more common than elsewhere in the world. Even in plan, the curve is more frequently used in Latin America." Hitchcock concluded: "A certain lyricism—of which color and curved forms are both important ingredients without being by any means universal—seems to have a continuous appeal to the Iberian temperament."[12]

Hitchcock's definition of Latin America from outside the region is framed as an architectural historian's search for formal resonances; by contrast, for Francisco Bullrich, writing the first history of Latin American modernist architecture from within the region and as a practicing architect, the question of character and identity had a deeper significance. In fact, Bullrich presented two versions of his history, both in 1969—*Arquitectura latinoamericana, 1930–1970* for a Spanish-speaking audience, as well as *New Directions in Latin American Architecture*.[13] While Hitchcock, in accordance with his internationalist perspective, had grouped the buildings by type with national origin duly noted among the specifications, in *Arquitectura latinoamericana* Bullrich grouped them by country, emphasizing the importance of local specificities—a decision that, in a sense, demonstrated more strongly the limits of the exercise, necessitating a justification for the selection of some countries and the absence of others, such as Colombia. The *New Directions* version had fewer chapters dedicated to individual countries and an increase in transnational thematic comparisons, under headings such as townscape, technology, and monumental architecture. These organizational changes reflect the different audiences of the period (or at least Bullrich's idea of them)—for example, the decision to present Villanueva's work to an English-speaking audience through a dedicated chapter rather than under the heading "Venezuela" obviated the need to contextualize the country's history, allowing Bullrich's focus to remain on the architecture itself.

In his introduction—which appeared with some differences in both editions under the title "Past and Present"—Bullrich emphasized that the diversity of architectural production in the Pre-Columbian period had resulted in very different levels of receptiveness to the project of incorporating its forms and motifs into contemporary architecture. Further, the Spanish colonial practice of establishing key urban centers over the ruins of conquered civilizations tended to concentrate the major works of colonial architecture in these same locations, in the process creating some unexpected continuities between the cultures. As a result of these settlement patterns, Bolivia, Mexico, and Peru were immersed in great colonial architecture, "in which the artistic tendencies of the European colonizers were integrated with the native craftsmen's needs for expression (or with the African's experience, as in the case of Brazil)";[14] smaller centers, lacking the same level of professional sophistication and material investment in architecture, had inherited a far more modest legacy.

For Bullrich, an authentic regional expression must take note of these historical factors, without slipping into the anachronistic approach of the neocolonial architecture so prevalent in the first decades of the twentieth century. On the other hand, Rationalist modernism as it had appeared in Latin America in the 1920s tended to overemphasize internationalism, rejecting as retrograde any suggestion of local character. Bullrich argued that this conflict between regionalism and cosmopolitan universalism, between tradition and modernity, could only be resolved through an engagement with "function" understood more broadly, in its cultural, locally specific dimensions. The dialectical development of this "regionalistic diversification" was an ongoing, open-ended process, not a search to uncover an already-constituted essence, for this was "not a historical constant but a variable, not an impersonal force but a quality residing in the individual. It is the single work of art that contributes to bringing into being such a concept."[15] This character could not be divined by "the mere transcription of some objective regional facts," nor through a supposed biological inheritance (Mexican architects were not naturally connected to "the eternal Aztec racial instinct"); rather, Bullrich argued, national or continental expressions could only result from "a genuine creative process."[16]

While Bullrich proposed an elusive and ever-evolving "regionalistic diversification," the broad-brush formalist characterization emphasized by Hitchcock has often had greater weight, effectively consolidating an image of curved and colorful "free-form" modernism as the region's signature contribution. The selection of Villanueva and Niemeyer as stand-ins for the region, noted above, is consistent with this approach. Specifically, the central ensemble at Brasília translates Niemeyer's expressive vocabulary into a necessarily more portentous idiom, but retains some of its sculptural qualities, both in solid, quasi-Platonic shapes (oblong towers and shallow domes), and in light, baroque gestures (the curling chapel spire, or elaborate staircases). The Palace of the Dawn achieves its iconic status by synthesizing these sculptural qualities with the image of a Greek temple conveyed by the frame of slender, tapering white columns—the image of democracy transposed, modernized, re-established in a new nation. The case of Villanueva's Aula Magna is somewhat more paradoxical, in that this undeniably iconic image is not particularly representative of Villanueva's work as a whole, and the striking biomorphic forms—colorfully curving and thus archetypically "Latin American," according to Hitchcock's definition—are the contribution of US sculptor Alexander Calder. Nonetheless, both projects can be read through Hitchcock's viewpoint as expressing the region's "lyricism" and formal bravura.

While conceding the iconic status of these projects, we suggest a reading that avoids both Hitchcock's formalism and Bullrich's emphasis on "regionalistic diversification" and its implied adjudications between sets of opposing terms (tradition and modernity, local and universal). Rather, we argue that using a transnational comparative approach—incipient for example in later attempts to compare Brasília and Le Corbusier's Chandigarh—reveals the complexities inherent in the postwar expansion of modernism to new global contexts. Similarly, the Aula Magna's interior, as Bullrich observes, represents the high-point of Villanueva's interest in the synthesis of the arts, and is thus indicative of the international reach of such architectural debates, even into "marginal" locations. In this reading, the images open up a network of transnational connections.

While Hitchcock and Bullrich have provided the foundations for subsequent debates, efforts to compile and examine the modern architecture of the region in its entirety have a longer and more complex history—not least because "modern" has not always been defined as modernism.[37] As a case in point, one of

INT.04 Alvarez, *Las obras de arquitectura*: Railway Station and streetscape, Panama

INT.05 Alvarez, *Las obras de arquitectura*: Central Post Office, Santiago, Chile; Office Building, Memphis, Tennessee; High Court Building, Caracas, Venezuela

the earliest surveys, *Las obras de arquitectura en la América Latina y en los Estados Unidos de América* (1921) by Mexican architect and engineer Manuel F. Alvarez, mobilized the eclectic classicism then in vogue to prove the decidedly European nature of a modern Latin America—a modernity it shared equally with the United States, as represented by projects from New York, New Orleans, and even Memphis. Guided by the photographs and information he could gather from the Pan-American Union in Washington DC, Alvarez presented the material country by country, moving south from Mexico and then ending with the United States. His intermittently comparative narrative, full of historical notations, focused on selected building typologies, such as presidential palaces, parliament buildings, and banks, depicting a Beaux-Arts universe of elite power to demonstrate the progress which the region—south and north—had achieved. Without completely rejecting colonial architecture, and attacking the prevailing tendency toward "utilitarianism," he endorsed the universal validity of the Beaux-Arts as sign of modernity, arguing that "the new architecture needed nothing of the aboriginal."[18] However, despite this claim, an anxiety over cultural identity and origins permeated the region's architectural production—Alvarez's endorsement of cosmopolitan academicism represented a continuation of Porfirian aesthetic values within revolutionary Mexico, but only a year later Mexico represented itself in a completely different guise, employing a decidedly colonial-style pavilion at the 1922 Centennial Exhibition in Rio de Janeiro.[19]

Panoramic histories such as Alvarez's—if we may call it so, since it more accurately falls under the category of operative documentary surveys—are rare. In the early twentieth century, transnational overviews were generally ushered in under culturalist calls from an insurgent regionalism, such as Argentine architect and theorist Angel Guido's *Orientación espiritual de la arquitectura en América* (1927, developed for the Third Pan-American Congress of Architects) which claimed neocolonial architecture as a synthetic art that best expressed the character of a modern "American" civilization. By the second decade of the twentieth century, linguists had noted an increase in the use of the term *América latina*,[20] which was perceived early on as an "infection." The 1921 Segundo Congreso de Historia y Geografía Hispano-Americanas, held in Seville, officially rejected any derivation of the term, underscoring that the only correct usage was *América española* (Spanish America) or *Hispano-América*.[21] This carefully circumscribed the category to Spanish-speaking nations—excluding Brazil and Haiti, among others—underscoring the cultural and political legacy of Spanish colonization in *América*. The Great War had initiated the demise of French and British interests in the region and stimulated those of the United States; simultaneously, Spain, propelled by an economic renaissance blossoming out of its wartime neutrality, moved to reestablish its influence through renewed economic and cultural ties. At stake within the region was the sense of a shared civilization that at once expressed ambivalence toward its own colonial past, and also a sense of unity or "solidaridad Americana" (American solidarity) concerning the future.[22]

In this period of rapid change, the colonial era came to be imagined as a synthetic historical moment that had produced the fusion of three cultures—Iberian, Indian (or indigenous American), and African—under a single stylistic form, the baroque. According to this idealized history, the colonial era had captured the essence of a racial and cultural *mestizaje* (hybridity), which would form the basis of a unique new identity; it had also managed to establish a vast and diverse imperial geography that had nonetheless allowed specific modulations in response to local requirements, manifesting an incipient regionalism without descending into chaotic fragmentation. The result was an essential Latin American architectural language—at once cementing nationalist discourses while transcending national boundaries. The vision of a new racial formation played out differently according to the socio-political conditions of each context: the position of writers in the Southern cone, with "minute" indigenous populations, contrasted with those in revolutionary Mexico, most notably José Vasconcelos, who forwarded his views as Secretary of Public Education, sponsoring architectural projects such as Carlos Obregón Santacilia's neocolonial Centro Escolar Benito Juárez (1923). Vasconcelos foresaw the eventual elimination of racial distinctions by "raising" the Indian into a mestizo—but predominantly white and Spanish—"cosmic" race.[23] Within Mexico, this interpretation of *mestizaje* was contested by Manuel Gamio, who in *Forjando patria* (1916) forwarded the idea of shared cultural bonds while preserving a heterogeneous society.[24] Both views had a significant impact on the discursive development of *mestizaje*, which fuelled, among other artistic expressions, the pictorial imaginary of the influential Mexican muralist movement.

Efforts to develop a national style based on an idealized reconciliation with the past, through which elite cultural nationalists mined tradition and popular culture, were invested with a sense of political urgency across the region. The defenders of the neocolonial—and its predominant stylistic codification, the neo-baroque— saw themselves confronting the disintegration of "national" cultures, battling a stylistic eclecticism in architecture that paralleled a broader cultural confusion provoked by intensifying modernization: the increased mobility of mediated images, the arrival of cinema, the growth of a consumer middle class, the general- ized desire to embrace novelty. However, theorists did not agree on a singular interpretation of the relationship between architectural style and national character, between (European) origin and (local) adaptation, between the relative contributions of center and periphery. For Argentine architect Alejandro Christophersen, colonial architecture was simply an extension of Spanish production, but its value resided precisely in this direct continuity.[25] By contrast, for Martín Noel—an Argentine historian of the colonial, as well as an active designer of neocolonial architecture—the colonial was not a mere extension of metropolitan architecture, but a specific Creole adaptation that emerged from, yet was not dependent on, its Iberian origins. Noel's *Contribución a la historia de la arquitectura Hispano-Americana* (1921) remains exemplary, despite its evident ideological eccentricities, such as calling Mexico "la República Española del norte"

INT.06 Noel, *Contribución a la historia de la arquitectura Hispano-Americana*: Monastery of Santa María de Alcobaça, Portugal; Totora boat, Lake Titicaca— The dialogue between European form and indigenous craftsmanship

(the Spanish Republic of the north), and its limited scope, since it primarily examined the territory covered by the Viceroyalty of Peru, in an attempt to establish a clear genealogy for Argentina's own colonial architecture.[26]

In Brazil, similar concerns over the powerfully transformative forces of modernization once again fuelled the emergence of the neocolonial, although in this case, with no other Portuguese settlement in the Americas to allow a regional comparison, neocolonial tendencies were generally framed within a Luso-Brazilian (metropolis-periphery) dialogue. Figures like Ricardo Severo, a Portuguese engineer and man of letters exiled in São Paulo, and José Mariano Filho, a Rio de Janeiro doctor, represent the complexity of the dialectic between origin and adaptation. As Jorge Schwartz has pointed out, in the continentalist rhetoric that surrounded the term *América latina* in the 1920s, Brazil stood as an outsider protected by the Tordesillas line—maintained by a language barrier—but this line was definitively crossed by the neocolonial, whose architectural vocabulary was flexible enough to incorporate various local inflections while still speaking of a common Iberian cultural experience and historical legacy in *América*.[27]

In no other buildings were concerns over national representation and indigenous modernity[28] better expressed than exposition pavilions. The 1929 Ibero-American Exhibition in Seville (in architectural studies always overshadowed by Mies van der Rohe's contribution to the International Exhibition in Barcelona) helped to forward the cause of the neocolonial and its *Hispanismo* ideology within an expanded geography of *Iberoamérica* that included Portugal and Brazil as participants, along with the United States. The United States pavilion, designed in a neocolonial style by California architect William Templeton Johnson (and including a stand-alone movie theater), helped to underscore the decidedly modern, adaptive character of neocolonial forms. The exhibition amply represented the colonial heritage (Portugal even built a pavilion for its colony in Macau), but also divergent cultural expressions, such as Manuel Ámabilis's Mexican Pavilion, inspired by Toltec-Maya forms imbued with socialist overtones of class–race unity, and the Colombian Pavilion by Spanish architect José Granados de la Vega, infused with indigenous motifs by Colombian artist Rómulo Rozo.[29] While the neo-baroque remained the predominant stylistic expression, this blending of indigenous and Art Deco influences broadened and shifted the notion of historical citation, and pointed to an aesthetic fusion that brought art and architecture closer to the intellectual preoccupations over *mestizaje*.

Despite such experiments, the celebration of baroque and indigenous forms within the general framework of the neocolonial remained an insurgent position, confronting an entrenched Beaux-Arts academicism continually reinforced in each new generation through architectural education. The predominance of neocolonial buildings at the 1922 Exposição Internacional do Centenário da Independência in Rio de Janeiro had represented a significant weakening of Beaux-Arts hegemony, in sharp contrast to the 1908 Exposição Nacional in Rio, or the 1910 celebrations for the centennial of independence across Latin America, which had resulted in an array of classicized public buildings aimed at enhancing cosmopolitan modernity, such as the Museum of Fine Arts or the Central Post Office in Santiago, Chile, the National Theater (later Palace of Fine Arts) in Mexico City, or the Monument to the Roman goddess Minerva in Guatemala City.

In 1922, concurrently with the official celebrations in Rio, yet another perspective on modernity unfolded: Modern Art Week in São Paulo adopted a quintessentially avant-garde stance, coalescing radical positions in the visual arts, architecture, literature, poetry, and music, to direct a series of affronts and shocks at the local bourgeoisie and the ruling oligarchy of the "Old Republic"—efforts which must be considered in relation to experimental gestures by artists and intellectuals across the region, such as the Estridentista Movement in Mexico.[30] In architecture, the emergence of modernist avant-gardes was accompanied by the gradual yet decisive rejection of both the Beaux-Arts and the neocolonial; the new Functionalist aesthetic of industrial internationalism challenged notions of regionalism and *mestizaje*, tending to eliminate signs of cultural specificity. In this way, the relationship between identity, modernity, and architectural form was further complicated: the adaptive character of modernism offered the potential for local inflections—the "regionalistic diversifications" prized by Bullrich—but these could easily become codified. In the case of Brazil, for example, the notion of an intrinsic sensuality became enmeshed with a "tropical" idiom, promoting the idea of a modern, national architecture that celebrated but at the same time masked the realities of cultural and racial difference. As sociologist Gilberto Freyre's *The Masters and the Slaves* (1933) revealed—and more poignantly, the poetry of Nicolás Gillén in Cuba—the African legacy would be even more difficult to incorporate than the indigenous cultures that, in a very selective manner, had been drafted to further the culturalist national projects of the 1920s.

Geographies of Modernism

By the 1930s, widespread but sporadic experiments with Functionalism and other modernist idioms had taken place throughout the region, and the theoretical basis of the new architecture was being articulated through speeches, articles, and manifestos. In Brazil, Gregori Warchavchik's "Apropos of Modern Architecture" (1925) and Lucio Costa's "Reasons for the New Architecture" (written 1930, published 1936) led the charge for the emerging modernists, while Flávio de Carvalho's "City of Naked Man" (1930) represented a point of connection to the broader avant-garde, specifically Oswald de Andrade's "Anthropophagic Manifesto" (1928), which advocated a creative "cannibalism" enabling local groups to appropriate foreign sensibilities for their own expressive purposes.[31] In Mexico, the *Pláticas sobre arquitectura* (1933) revealed the tensions in Mexican architectural culture by assembling a diverse group of architects, including Juan O'Gorman, to articulate the contributions (and limitations) of Functionalism in the context of post-revolutionary nation-building, while José Villagrán Garcia's "Apuntes para un estudio" (1939–1940) explored the fundamentals of the new architecture from the viewpoint of one of its most influential teachers.[32] Villagrán García had begun teaching architectural theory at the National University in 1927, as discussions on Functionalism were carried beyond academic circles and into public consciousness through newspapers such as *Excélsior* in Mexico City, *La Prensa* in Buenos Aires, and *Correio da manhã* in Rio de Janeiro. Elsewhere vanguardist groups issued manifestos claiming affinities with an international Modern Movement while grounding themselves in the specificities of local conditions, often emphasizing the architect's social responsibility—this was the case with Grupo Austral in Argentina (1939), ATEC in Cuba (1943), and Agrupación Espacio in Peru (1947),[33] all of which were connected to CIAM in one way or another.

On the other hand, there were—and still are—few general histories of architecture that include examples of modernism in Latin America. When European modernists presented the first surveys of the "new international architecture" their gaze rarely strayed from Europe and the United States. Walter Gropius's *Internationale Architektur* (1925) included photographs of grain silos in Argentina, alongside those in the United States and Canada, manifesting a territory of pure functional structures in which European architects could find creative inspiration; the only architectural project per se to reference the region was Hugo Häring's 1923 proposal for the German Club in Rio de Janeiro, suggesting the intricate

La industria moderna se exije perfección mecánica
y formas resultantes estéticas.

INT.07 Jose Villagrán
García, "Notes for Study III:
Man"—The perfection of
modern industrial forms
facilitates health and
beauty

La juventud de hoy
exije para sí formas
perfectas, salud y be-
lleza.

connections of immigration and commerce between Latin America and non-"Latin" Europe.[34] Alberto Sartoris's *Gli elementi dell'architettura funzionale* (1932) was perhaps the first survey to introduce modern works by Latin American architects to European readers, proving the incompleteness of Henry-Russell Hitchcock and Philip Johnson's 1932 survey of International modernism at MoMA. Although coverage in the first edition was limited to the work of Gregori

INT.08 Gropius,
Internationale Architektur:
Hugo Häring, Project for
German Club in Rio de
Janeiro, 1923

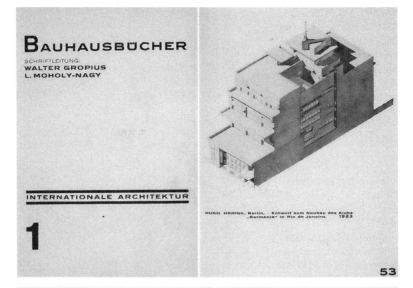

INT.09 Sartoris, *Gli
elementi dell'architettura
funzionale* first edition:
Gregori Warchavchik,
Modernist House, São
Paulo, 1930

INT.10 Sartoris, *Gli elementi
dell'architettura funzionale*
second edition: León
Dourge, apartment
building, Buenos Aires,
1934; Mauricio Cravotto,
Villa and Studio in
Montevideo, 1933

Warchavchik in Brazil, such as his Modernist House (1930), in the 1935 edition Sartoris expanded the survey to incorporate examples in Argentina and Uruguay, such as Wladimiro Acosta's Integral-Block City project (1931), Antonio Vilar's Club Hindú (1932) and León Dourge's Apartments (1934), both in Buenos Aires, and Mauricio Cravotto's Villa and Studio in Montevideo (1933).[35] For Sartoris—as for Gropius and other partisans of the new architecture—the aim was to document an international modernist language, with no concern for "regionalistic diversification" as identified by Bullrich; Latin America provided an additional breadth of territorial coverage as proof of Functionalism's universal reach and validity.

Le Corbusier (who wrote the preface to Sartoris's book) published *Précisions* (1930) as a record of sorts of his 1929 visit to the Southern Cone, reproducing the lectures he had given in Buenos Aires along with his thoughts on Brazil and the United States (as well as Paris and Moscow), but he had little to say about the architecture that he encountered, other than intimating the latent sensuality of its tropicalism.[36] By contrast, Werner Hegemann, whose visit to the region two years later was sponsored by a rival faction aiming to temper Le Corbusier's influence,[37] wrote a series of articles presenting a vivid account of the local scene. His subjects ranged widely: social housing projects in Buenos Aires built by the state and by Catholic agencies;[38] the "Schinkelesque spirit" of simple, "cubist" dwellings in Montevideo—"small machines for living in" built by untutored contractors and architects alike;[39] and the contrast between cramped inner-city tenements in Buenos Aires and self-built homes on the urban periphery—a scene that reminded Hegemann of 1930s Berlin.[40]

US architect–photographer Esther Born was an early eye-witness to the achievements of Mexican modernism: in the early 1930s the work of Juan O'Gorman, Luis Barragán, and Carlos Obregón Santacilia had been published in US architectural journals, reaching a highpoint with the theme issue on "The New Architecture in Mexico" that she edited for *Architectural Record* in 1937. Focusing on Mexico City, Born noted that the quantity of modernist architecture "would be unexpected in any North American city; but to the Northerner, acquainted with Mexico only through literature and hearsay, the energy displayed and the up-to-the-minute quality are doubly astonishing."[41] With examples of

INT.11 Born, "The New Architecture in Mexico": Juan O'Gorman, Escuela Técnica Tres Guerras, 1932 (Reprinted with permission from Architectural Record © 1937, The McGraw-Hill Companies. www.architecturalrecord.com)

single-family dwellings and workers' housing, luxury apartments and industrial structures, schools, monuments, office buildings, and shopping centers, the modernist idiom had permeated all aspects of urban life—adopted not just by wealthy patrons, but also by the public and private sector alike.

The Brazilian pavilion at the 1939 New York World's Fair—designed by Lucio Costa and Oscar Niemeyer, with Paul Lester Wiener—confirmed the rising importance of the region's modern architecture, which had often been heralded but few had seen.[42] By the immediate postwar period, its architectural production had reached such a high level of development and sophistication that it demanded broader international attention. MoMA's 1943 *Brazil Builds* exhibition led this shift, and through its initiative the Ministry of Education and Health (1936–1945) designed by a team of architects led by Lucio Costa became the period's prime icon of Latin American modernism.[43] Framed as a survey of architecture "1652–1942," the exhibition grouped early examples—rustic houses, baroque churches, colonial administrative buildings—by location, as if describing a touristic itinerary, but grouped modern buildings by type, emphasizing the breadth and variety of production.

If MoMA's highly influential exhibition helped focus interest on Latin American modernist architecture, it was not alone in its admiration. Sigfried Giedion's *A Decade of New Architecture* (1951), produced under the auspices of CIAM, included numerous examples from the region, such as the work of Nicolás Arroyo and Gabriela Menéndez in Cuba, Jorge Ferrari Hardoy and Juan Kurchan in Argentina, and Gabriel Solano and Jorge Gaitán in Colombia. In his introduction, Brazil was one of the two countries that Giedion singled out as expanding the global geography of a "democratic" modernism, "to the very fringes of our civilization in the north and in the south. The constructions that come from the soil of *Finland*

INT.12 Philip L. Goodwin,
*Brazil Builds: Architecture
New and Old, 1652–1942*
(New York: The Museum of
Modern Art, 1943). © 1943
The Museum of Modern Art,
New York

and of *Brazil* are not frontier enterprise or provincialism but works of inspiration and of new discovery."[44] The gaze toward the south reflected Giedion's interests as head of CIAM, just as he had deployed Josep Lluís Sert as an unofficial ambassador to expand Latin American membership.[45] By contrast, the new generation of CIAM's successor organization Team 10 would later turn inward with an almost exclusively Eurocentric membership and outlook.

Into the 1950s, architecture from across the region, particularly Brazil, appeared consistently in major architectural journals around the world. However, by the middle of the decade, critiques of the Brazilian approach were beginning to appear. In 1950, *Architectural Review* already intimated some doubts: it argued that international interest in Brazil was symptomatic of a "quest for novelty"; while low-cost housing and "social improvements are neglected," resulting in an "incomplete architecture" that had failed to benefit all levels of society.[46] Far more damning was the magazine's 1954 "Report on Brazil," comprising the assessments of five visitors to the recent São Paulo Bienal. While Walter Gropius was mildly critical (the Ministry building was "too dirty and has a few leaks"), Swiss designer Max Bill warned about the excesses and capricious licence of Brazilian modernism, its propensity for a formal exhibitionism that led to "anti-social academicism" and "utter anarchy in building, jungle growth in the worst sense." Ernesto Rogers was inclined to be somewhat more forgiving, blaming the architecture's "overbearing novelty" on the immoderation of the country itself, which—like its women—was "over-perfumed, over-colored, highly sensual."[47] Clearly directed at the work of Niemeyer, these critiques apparently aimed to curtail the influence of these formal explorations of mid-century modernism.

Aside from these contemporary surveys, the first historical account of modern architecture to incorporate the region was Bruno Zevi's *Storia dell'architettura moderna* (1950). Zevi included only brief references under generalizing terms, focusing on Brazil and Mexico. His most lasting effect was to subsume the region under the shadow of Le Corbusier, who in "designing for Latin America called forth the Brazilian renewal"; Le Corbusier's lessons and "themes," Zevi concluded "were rigorously applied . . . often in a scholastic manner."[48] Through such arguments, Zevi initiated the attempts to incorporate Latin American modernism into a European intellectual and cultural framework, tracing a Western universal heritage that saw the region as derivative and dependent on metropolitan cultural centers.

To a certain extent, Henry-Russell Hitchcock's *Architecture: Nineteenth and Twentieth Centuries* (1958) continued this construction. However, it remains unparalleled for its depth and nuance in integrating the region into the history of Western modern architecture. Hitchcock presented a complex yet problematic tapestry of interrelated examples: Le Corbusier's influence was decisive, for example, yet elsewhere Hitchcock intuited a "more emotive line of development in modern church architecture" that connected Antonio Gaudí and Dominikus Böhm to recent works by Oscar Niemeyer in Brazil, by Juvenal Moya in Colombia, by Enrique de la Mora and by Félix Candela in Mexico.[49] Despite his often negative assessments, in Hitchcock's view Latin American modern architecture offered a clear continuity of Western civilization, while demonstrating independence and maturity; consonant with W. W. Rostow's stages of economic development, it had indeed "taken off."[50] Hitchcock's historiographic project here paralleled MoMA's *Latin American Architecture since 1945*, which he had helped to organize; a decade after *Brazil Builds*, the region was booming despite the darkening clouds of the Cold War.

Leonardo Benevolo's *Storia dell'architettura moderna* (1960), which was highly influential in Latin America, was the first to contract the geography of the region's

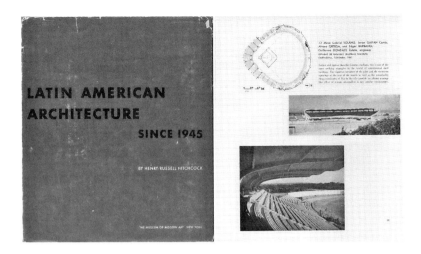

modernism to the singular example of Brazil.[51] Although for Benevolo the Brazilian experience signaled a shift in the cultural centers of modernism, and Brasília marked an important contribution that would influence later developments, in subsequent revisions of the text the Italian historian determined that this modernist experiment had been definitively interrupted by dictatorships and revolution. Sigfried Giedion echoed this narrative in his introduction— "Architecture in the 1960s: Hope and Fears"—to the 1962 fourth edition of *Space, Time and Architecture*. The region's contribution to modern architecture, embodied by Brazil's "glittering facades and astonishingly impressive projects," was now (unlike peaceful and "strongly democratic" Finland) imperiled by the menace of political upheavals.[52] Giedion's oblique reference to the Cuban Revolution reveals a shift in emphasis away from celebrations of formal bravura, toward a greater consideration of the social–political context of the works—which, however, often led to their easy dismissal as compromised or inadequate. As both postwar optimism and the "heroic" decades of Latin American modernism were coming to an end, this reconsideration became unavoidable. In 1976, Manfredo Tafuri and Francesco Dal Co's *Architettura Contemporanea* presented the architecture of Carlos Raúl Villanueva in Venezuela and Mario Pani in Mexico as elegant yet futile attempts to counter the architecture of bureaucracy that, following the lead of the US firm of Skidmore Owings and Merrill, had established itself globally. In a world dominated by corporate capital, the Italian historians concluded, the Brazilian neo-expressionist experiment had ended with Brasília, creating "spectacles of the absurd, euphoric fragments of crystallized nature."[53] Six years later, the 1982 Spanish edition of Benevolo's history included a chapter by Josep María Montaner dedicated to modern architecture in Latin America, which underscored the developmental conflicts of its inescapable Third World geography.[54] Such discord, Montaner pointed out, made its cities pass from freshness to decrepitude without ever being old—an observation Claude Levi Strauss had made decades earlier in *Tristes Tropiques* (1955). Through such accounts, the narrative of a derivative modernism embedded within the geography of dependency was firmly installed.

A possible exit from this impasse was opened up by Alexander Tzonis and Liane Lefaivre, who coined the term "Critical Regionalism" in 1981, emphasizing the place-specific dimensions of architecture—its responsiveness to climate, topography, local materials, and appropriate technologies. The concept had particular significance for Kenneth Frampton, representing the potential for "an architecture of resistance" to counter the homogenizing forces of a "universal" culture disseminated with the expanding outreach of global capital.[55] While the first edition of Frampton's *Modern Architecture: A Critical History* (1980) reiterated a familiar vision of Latin America, limiting the focus to Costa and Niemeyer in Brazil, in the preface to the second edition in 1985, Frampton introduced Critical Regionalism as a "'revisionist' form of modern architecture [that] has been in existence for the past forty years or more," singling out Argentine architect Amancio Williams's Bridge House (1945) as a key precursor of the tendency, along with Barragán's domestic architecture and Villanueva's Ciudad Universitaria.[56] In the fourth and most recent edition (2007) Frampton's view of the region is at once sharper, and more diffuse, in keeping with the loose thematic organization of the new sections, which deploy a wide range of examples to focus on Frampton's current core concerns—including topography, sustainability, materiality, and habitat. At several points, Frampton suggests an emerging historical consciousness of the traditions of Latin American modern architecture; for example, connecting Alejandro Aravena's Elemental Housing Project in Chile (2004) to the PREVI experimental housing estate in Peru (1969–1976).[57] Elsewhere Paulo Mendes da Rocha's work evokes not just "the heroic Brazilian tradition in reinforced concrete," but "an even broader Latin American will-to-build" incorporating both Williams and Dieste.[58] For Frampton, Latin America represents immense possibility and innovation; at the same time, the extreme income inequality long seen as a fatal flaw in Latin American societies has now permeated "a world that is increasingly divided by prodigious wealth and abject poverty";[59] rather than modernism foreshadowing the region's economic development, it appears that Latin America has foreshadowed the global phenomenon of inequality.

The latest effort to focus specifically on the region's modern architecture is Valerie Fraser's *Building the New World: Studies in the Modern Architecture of Latin America, 1930–1960* (2000). Fraser concentrates on paradigmatic examples from three countries—Mexico, Venezuela, and Brazil—largely as discrete and independent objects of study, although general notions such as "idealistic architects," "ambitious governments," or "showpiece projects," all deployed in the introduction, hint at transnational parallels which the reader is invited to imagine. The drive to compare the architectural production of the three countries tends to reinforce traditional transnational actors such as Le Corbusier, but also introduces others less well known, such as Richard Neutra. Fraser accepts the term "Latin America" as a signifier, without challenging the concept itself or critically articulating moments of its construction. This ambitious and eloquent study reveals the difficulties of the enterprise: the instability of the national frame, put into crisis by modernist architects attempting to create cosmopolitan works; the inescapable material conditions which fracture the national frame into a regional and cultural mosaic; and the complex forces within transnational imaginations that demand that the region's architecture perform in a cohesive, singular, and quasi-oppositional manner. As Fraser tells her readers: "Latin American modern architecture . . . is not an uncritical reworking of European modernism . . . but a deliberate and more profound adaptation of or challenge to European models."[60]

Within the region, efforts to support research into the architectural history of Latin America reached an important benchmark with Mario J. Buschiazzo's creation of the Instituto de Arte Americano e Investigaciones Estéticas in Buenos Aires in 1946. This expanded to a continental dimension the work fostered by Manuel Toussaint at the Instituto de Investigaciones Estéticas of the Universidad Nacional Autónoma de México since 1936, and by Martín Noel at the Laboratorio de Arte Americano in Seville, which developed out of the 1929 Ibero-American exhibition.[61] These efforts concentrated on the achievements of the colonial period—perhaps not surprisingly, since as modernism emerged in the 1940s colonial architecture came under threat: urban redevelopment projects such as Cipriano Domínguez's Centro Bolivar in Caracas, or Town Planning Associates' *Plan Piloto* for Havana, which proposed leveling historic urban cores, revealed the fragile condition of the past (as well the political alliances between modernism, developmentalism, and dictatorships in such contexts). Research institutes and other academic initiatives performed much-needed documentation of colonial architecture, and activated a network of Ibero-American scholars, reconnected to Spain as evidenced by Diego Angulo Iñiguez's *Historia del arte hispano-americano* (1945), or the encyclopedic twenty-two-volume *Ars Hispaniae*.[62]

In 1967, the Centro de Investigaciones Históricas y Estéticas (CIHE) of the School of Architecture of the Universidad Central de Venezuela gathered art and architectural historians from the Americas and Europe to assess the state of architectural history in the region, concluding that there was a need to extend research into the nineteenth and twentieth centuries, and more importantly, to unify the fragmented state of the discipline.[63] The conference also called for an active historiography that would connect the historian's endeavors to the contemporary problems of Latin America as a whole. This notion of an active historiography emerged from the center–periphery model of 1960s Dependency Theory, conceiving culture as the result of an ontological determinism that limited its diversity and plurality, at times negating its historical specificity. In this vein, Rafael López Rangel's *Arquitectura y subdesarrollo en América Latina* (1975), examined architectural production within the framework of underdevelopment, taking aim at supra-historical narratives based on ontological notions such as "the spirit of the nation" which he believed had plagued Mexican, and Latin American, historiography.[64] Roberto Segre's *América Latina en su arquitectura* (1975), published by UNESCO, took a different tack, initiating efforts to single out key "characteristics" of the region by assembling and giving voice to its "experts."[65] This approach has continued to guide endeavors such as Damián Bayón's *Panorámica de la Arquitectura Latinoamericana* (1977), which presented interviews with Eladio Dieste, Rogelio Salmona, Roberto Burle-Marx, and Clorindo Testa, among others.[66]

More recently, CIHE's initiative has been taken up by the Seminario de Arquitectura Latinoamericana (SAL) launched by the Argentine journal *Summa* in 1985. *Summa*'s calls for a new "appropriate" historiography were predicated on ending the derivative relationship between Latin American cultures and their metropolitan centers.[67] Ramón Gutiérrez's *Arquitectura y urbanismo en Iberoamérica* (1983) and Enrique Browne's *Otra Arquitectura en América Latina* (1988) were perhaps the most salient contributions to these efforts.[68] While such Spanish-language regional surveys invariably include Brazil, no histories of Latin American architecture have emerged from that country. The Tordesillas line remains quite present in Brazilian architectural scholarship, the national frame remaining the de facto organizational and discursive structure. Yves Bruand's *Arquitetura contemporânea no Brasil* (1981) remains the benchmark of this approach.[69] Recent efforts such as *Urbanismo na América do Sul* (2009) attempt to address this important lacuna in Brazilian historiography by underscoring transnational debates and focusing on institutional frameworks such as the Pan-American Congress of Architects, or the trajectories of professionals who were active in a macro-regional context, thereby giving shape to a broader, shared "South American" architectural and planning culture.[70]

Given the difficulties of constructing "continental" histories, in particular the challenge of avoiding essentialist claims, it is understandable that recent histories of Latin American modern architectures have tended to focus on the national frame. However, architectural history is only beginning to unravel the complex relationships between modernism and the processes of nation-making at the core of modernity, and at times the convenience of the national frame has discouraged a critical examination of the slippages between state and nation, culture and identity, and most importantly, representation and agency in elite versus more inclusive constructions of the nation. Today the mobilization of architectural modernism under the banner of preservation and patrimony, as advocated by Docomomo, recalls the mandate of organizations formed in the first decades of the twentieth century to preserve the colonial heritage which the rise of modernism threatened to destroy. In both cases, the documentation of, and advocacy for, physical structures has been invaluable, but often lacks a sufficiently nuanced cultural analysis of the histories surrounding them.

Nonetheless, the scholarship on the region's architecture is—contrary to what many Anglophone readers believe—broad and diverse. In a by-no-means exhaustive survey of notable recent studies, Fernando Pérez Oyarzún's *Le Corbusier y Sudamérica* (1991) must be singled out as a critical moment in the re-examination of the architect's relationship with Latin American architecture, with a collection of essays exploring the transnational connections between Brazil, Argentina, Uruguay, and Colombia forged through his multiple visits to the region.[71] Elsewhere, the interest in a larger and more complex cast of characters is signaled by studies such as José Lira's recent *Warchavchik: fraturas da vanguardia* (2011) which combines a catalogue raisonné with efforts to recreate the multiple cultural contexts the architect inhabited as an immigrant from Odessa to Brazil, via Italy.[72] Similarly, Alberto Sato Kotani's monograph *José Miguel Galia, arquitecto* (2002)— an Argentine who trained in Uruguay under Julio Vilamajó and later moved to booming 1940s Venezuela—reveals the fluidity and complexity of a cosmopolitan Latin American modernism.[73] In another register, research into historical avant-gardes has produced valuable collections of primary documents (by Enrique X. de Anda and Salvador Lizárraga in Mexico) and of critical scholarship (by Abilio Guerra in Brazil).[74]

Other historians have presented portraits of groups of architects: Jorge Francisco Liernur's *La red austral* (2008) begins with a discussion of Argentine debates of the 1920s that paved the way for the reception of Corbusian ideas, and then explores how these played out in projects by local "disciples" up into the 1960s.[75] Concentrating on the pivotal decades of mid-century Brazil, Lauro Cavalcanti's *As preocupações do belo* (1995) examines a series of government projects, expanding the focus from the canonical Ministry of Education and Health to contemporaneous but stylistically contrasting ministry buildings in the same Rio district, thereby illuminating their shared role within the nationalist–developmentalist project of Getúlio Vargas's Estado Novo.[76] John A. Loomis's *Revolution of Forms* (1999) focuses on the group of architects charged with design of the National Art Schools in Havana at a critical moment of the Cuban Revolution.[77] Silvia Arango's *Historia de un itinerario* (2002) considers university campuses in Bogotá and Caracas, as well as the Cuban art schools, to forward the "ciudad universitaria" as a typology of particular significance in Latin American modernism—an observation also made by Henry-Russell Hitchcock in the 1950s— and thereby opens the way for further transnational comparisons with similar projects in Mexico, Panama, and elsewhere.[78] The rediscovery of the Cuban art schools has prompted regional surveys aiming to preserve the modernist heritage in the Caribbean; of these, the recent collection on Henry Klumb's work in Puerto Rico remains notable.[79]

Luis E. Carranza's *Architecture as Revolution* (2010) employs case studies of state-sponsored neocolonial and later Functionalist buildings to elucidate the complexity of the building culture of post-revolutionary Mexico; the inclusion of the avant-garde Estridentista Movement highlights important connections to the cultural and sociopolitical milieu, often forgotten in architectural scholarship.[80] Anahí Ballent's *Las huellas de la política* (2005) presents a nuanced view of the intersections between politics and architecture within the very different context of Peronist Argentina, and is one of the few studies to include a discussion of architecture produced on a commercial basis, in this case housing for a status-conscious, middle-class clientele.[81]

Nineteenth-century topics remain particularly understudied; in general, histories of urbanism have been more effective in drawing out the complexity of dialogues between Latin America and Europe in the decades around the turn of the twentieth century, and less committed to proving a modernist "break," emphasizing instead continuities and convergences. Arturo Almandoz's *Urbanismo europeo en Caracas* (1997) and Adrián Gorelik's study of Buenos Aires, *La grilla y el parque* (1998), are exemplary in this regard.[82] In a different vein, Alberto Saldarriaga Roa's *En busca de Thomas Reed* (2005) provides an illuminating study of a peripatetic English-trained architect and engineer, who worked in Colombia, Venezuela, and Ecuador, personifying the infiltration of British technological expertise into the region in the late nineteenth century.[83]

Finally, comparative transnational research, such as that on transatlantic modernism, has barely begun. However, Robert González's *Designing Pan-America: US Architectural Visions for the Western Hemisphere* (2011) has opened up a new territorial framework for exploration in an informative study that ranges from a re-historicized perspective on Chicago's World's Columbian Exposition of 1893 to the 1929 competition for the Columbus Memorial Lighthouse in Santo Domingo, which attracted an extraordinarily diverse group of entrants including Konstantin Melnikov, Tony Garnier, Carlos Obregón Santacilia, and Alvar Aalto.[84] Along these lines, hemispheric and international fairs, Pan-American Congresses, not to mention the Pan-American Highway, offer fertile ground for architectural historiography.

Ambiguous Territories

Our selection for this book has been based on the strengths of individual contributions, and on the range of topics, contexts, and figures that they consider, allowing us to present a rich survey of the field of Latin American modern architectures. The scholars themselves come from a diverse range of intellectual traditions, trained variously in Latin America, the United States, and Europe, and a secondary aim of our selection has been to initiate stronger transnational connections between these different traditions, in particular countering the prevalent disconnection between Spanish- and Portuguese-language scholarship, and that in English. Our grouping of the chapters into three sections has emerged from our perception of key underlying themes, which we have categorized as Singular Journeys, Techno-Cultural Assemblages, and Mediated Territories.

Singular Journeys presents significant figures of modernist architecture in non-canonical contexts, thereby complicating our understanding of their histories. Generally monographic in approach, the chapters address issues of transfer, influence, and exchange via the transnational networks established through the professional trajectories of individual careers. Gaia Piccarolo reframes a key figure in Latin American modernism, Lucio Costa, by examining his transatlantic search for the meaning of Brazil's colonial heritage, questioning the assumed unidirectionality of influence between metropolitan center and colonial periphery. Jorge Nudelman recontextualizes the role of Le Corbusier in the region by demonstrating that Germany, as well as France, figured prominently in Uruguayan architects' understanding of "European" modernism; Nudelman considers the case studies of architects who worked with Le Corbusier in Paris, viewing his impact through their individual experiences rather than accepting a generalized assumption of Corbusian influence. Noemí Adagio considers architecture's post-war humanist crisis in two contrasting contexts, finding resonances between the pedagogical projects of Joseph Hudnut at the "Harvard Bauhaus" and of Enrico Tedeschi—an Italian architect associated with the Metron group—in Mendoza, Argentina.

Central to the section on Techno-Cultural Assemblages are the issues of development, and of developmentalism—the whole-hearted commitment to overarching modernization projects launched by countries throughout Latin America under the centralist, top-down guidance of the state, in an effort to replicate the industrial revolutions experienced in Europe and North America. These programs were often complex and tightly coordinated, comprising economic, social, and spatial-territorial planning. Luis Castañeda examines the parallel careers of two architect-politicians—Fernando Belaúnde Terry in Peru and Pedro Ramírez Vásquez in Mexico—considering the fate of their efforts to create a local identity by integrating Pre-Columbian traditions into the construction of a modern industrial nation in projects ranging from roadways to museums. Viviana d'Auria looks at one of the key examples of unfettered Latin American developmentism—the massive social housing projects produced under the dictatorship of General Pérez Jiménez in Venezuela—following the transformations of these projects from the 1950s to the present in the context of their ever-evolving political, social, and urban environment. With Daniel Talesnik the theme of development is continued by focusing on one high-profile architectural project produced at a critical political and social juncture in Chile, presenting a vivid case study of a United Nations building constructed under Salvador Allende's short-lived socialist state, and symbolically re-constructed under the military regime that overthrew him. Pedro Ignacio Alonso and Hugo Palmarola Sagredo offer a contrasting perspective on the same politically charged period, highlighting the processes of development through the importation of prefabricated housing technologies from the Soviet Union into Chile, as well as Cuba, as part of an effort to consolidate Cold War transnational alliances. The final two chapters in this section deal with the question of development on a macro scale. Claudia Shmidt examines nineteenth-century debates over the formation of a modern nation-state through the case of Argentina, where controversy over the relationship between the capital city and the national territory continued unresolved for several decades after independence. Paulo Tavares re-frames the building of Brasília as a bridgehead for the large-scale resource-development and urbanization of the Brazilian Amazon, with particular focus on the developmentalist agenda of the military government installed in 1964 and its Cold War program of internal colonization; in the process, Tavares decisively extends the geography of Brazilian modernism beyond urban centers such as Brasília, Rio de Janeiro, and São Paulo.

Mediated Territories considers the intersections of national territories and imagined communities—to use Benedict Anderson's phrase—through representations of architecture in images and journals.[85] Cristina López Uribe examines the migration of popular forms of neocolonial residential architecture between Mexico and the United States in the 1920s and 1930s, through the mediums of film, advertising, and architectural magazines, demonstrating that the modernity of these buildings resides in their formation through this exchange and recirculation of images. Hugo Mondragón López compares two nations—Colombia and Chile—as they were imagined by two key architecture journals, elucidating the implications of modernization for architectural culture in these contexts. George F. Flaherty focuses on the figure of Mario Pani in his role as editor and publisher of *Arquitectura/México*, examining his construction and representation of Latin America through one of the region's most important architectural journals. María González Pendás draws out resonances between the political and the architectural concerns of Spanish émigré Félix Candela, studying his geopolitical theories as well as his production of hypar structures in Mexico; here the frame goes beyond Latin America to consider Candela's assemblage of an "Ibero-American" political entity as a strategy to deliver Spain from its repression under General Franco.

As we have noted, the category "Latin America" has often been critiqued as essentializing and limiting; however, the national frame presents its own limitations as the default option of architectural history. Rather, our aim in this collection has been to move away from narratives defined in terms of national boundaries and nationalist imperatives, by "narrowing the frame" to examine local histories or micro-histories, and "widening the frame" through comparative and transnational approaches. By insisting on this relational geography, we aim to emphasize the ambiguous, networked territories of modern architectures—a pattern that is exemplified here by Latin America, but is not unique to the region, and thus opens up an alternative model for emerging scholarship on other locations. To return to Banham's term we noted at the outset, one form of "continental failure" would be to imagine Latin American modern architectures as homogeneous, to assume their singular identity; it would be a failure of another kind to refuse to imagine Latin American histories altogether.

Notes

1 Reyner Banham, "Book Review: *New Directions* Series," *The Art Bulletin* 54, no. 4 (December 1972): 565.
2 Michel Chevalier, quoted in Arturo Ardao, *Génesis de la idea y el nombre de América Latina* (Caracas: Centro de Estudios Latinoamericanos Rómulo Gallegos, 1980), 29. This and all subsequent translations are by the authors.
3 For an in-depth examination of the term see Ardao, *Génesis de la idea*; also Dilma Castelo Branco Diniz, "O conceito de América Latina: Una perspectiva francesa," in *Anais do XI Encontro Regional da ABRALIC* (São Paulo: Abralic, 2007); Walter Mignolo, *The Idea of Latin America* (Malden; Oxford: Blackwell Publishers, 2005).
4 Nonetheless, as Erika Pani points out "the empire only effectively controlled the central regions of the country . . . and the road from the capital to Veracruz." Pani, "Dreaming of a Mexican Empire: The Political Projects of The 'Imperialistas,'" *The Hispanic American Historical Review* 82, no. 1 (2002): 23. See also Raymond B. Craib, *Cartographic Mexico: A History of State Fixations and Fugitive Landscapes* (Durham: Duke University Press, 2004).
5 The historical rehabilitation of the French mission began with Afonso de E. Taunay, *A missão artística de 1816* (Rio de Janeiro: Ministério da Educação e Cultura, 1956). See also Alberto Sousa, *Arquitetura Neoclássica Brasileira: Um Reexame* (São Paulo: Pini, 1994).
6 See José Enrique Rodó, *Ariel*, trans. Margaret Sayers Peden (1900; Austin: University of Texas Press, 1988); Roberto Fernández Retamar, "Caliban: Notes Toward a Discussion of Culture in Our America," in *The Latin American Cultural Studies Reader* (1971; Durham: Duke University Press, 2004), 83–99.
7 See Ricardo Donato Salvatore, *Imágenes de un Imperio: Estados Unidos y las formas de representación de América Latina* (Buenos Aires: Editorial Sudamericana, 2006); and Helen Delpar, *Looking South: The Evolution of Latin Americanist Scholarship in the United States, 1850–1975* (Tuscaloosa: The University of Alabama Press, 2008).
8 Sylvester Baxter, *Spanish-Colonial Architecture in Mexico* (Boston: J. B. Millet, 1901), 8.
9 Rachel Weiss and Alan West, eds., *Being América: Essays on Art, Literature, and Identity from Latin America* (Fredonia: White Pine Press, 1991), 123.
10 Carlos Brillembourg, ed., *Latin American Architecture, 1929–1960: Contemporary Reflections* (New York: Monacelli Press, 2004).

11 Henry-Russell Hitchcock, *Latin American Architecture since 1945* (New York: Museum of Modern Art, 1955), 16–17.

12 Hitchcock, *Latin American Architecture since 1945*, 26. In a manner reminiscent of Chevalier, Hitchcock also argued for the importance of Catholicism as "major element of cultural homogeneity" across the region; even though the Church may not be interested in modern architecture per se (the local bishop refused to consecrate Niemeyer's Church of São Francisco at Pampulha, for example), it has powerful cultural effects in creating large families which make apartment living impractical. Another general formal trait was created by the need for security which demanded that almost everywhere openings be "grilled or otherwise protected to keep out thieves."

13 Francisco Bullrich, *Arquitectura latinoamericana, 1930–1970* (Buenos Aires: Editorial Sudamericana, 1969); Bullrich, *New Directions in Latin American Architecture* (New York: George Braziller, 1969).

14 Bullrich, *New Directions*, 14.

15 Ibid., 19, 18.

16 Ibid., 17, 18.

17 Here we are focusing on panoramic surveys of the region, rather than local or national histories. For a wider, yet ideologically compromised, examination see: Ramón Gutiérrez, "La historiografía de la arquitectura americana: Entre el desconcierto y la dependencia cultural, 1870–1985 (1)," *Archivos de arquitectura antillana: AAA* 2, no. 3 (January 1997); "La historiografía (2)," *AAA* 2, no. 4 (May 1997); "La historiografía (3)," *AAA* 2, no. 5 (September 1997).

18 Manuel F. Alvarez, *Las obras de arquitectura en la América Latina y en los Estados Unidos de América* (Mexico City: Secretaria de Comunicaciones y Obras Públicas, 1921), 9.

19 See Mauricio Tenorio-Trillo, *Mexico at the World's Fairs: Crafting a Modern Nation* (Berkeley: University of California Press, 1996).

20 Aurelio M. Espinosa, "The Term Latin America," *Hispania* 1, no. 3 (1918).

21 A. M. E., "The Term Latin America Repudiated," *Hispania* 4, no. 4 (1921). Ramón Menéndez Pidal underscored that America could not be Latin, since this implied a direct inheritance from the ancient peoples of Latium in central Italy, as was the case with the French, Romanian, Portuguese, Sardinian, and Spanish cultures.

22 See Antonio Manero, *México y la solidaridad americana; La doctrina Carranza* (Madrid: Editorial-América, 1918).

23 See José Vasconcelos, *The Cosmic Race/Raza Cósmica: A Bilingual Edition*, ed. Robert Reid-Pharr, trans. Didier T. Jaén (1925; Baltimore and London: Johns Hopkins University Press, 1997).

24 Rick Anthony López, *Crafting Mexico: Intellectuals, Artisans, and the State after the Revolution* (Durham: Duke University Press, 2010), 131.

25 For an important recent reassessment of this architecture, see Aracy A. Amaral, ed., *Arquitectura neocolonial: América Latina, Caribe, Estados Unidos* (São Paulo: Fundação Memorial da América Latina, 1994).

26 Having been a second-tier settlement prior to becoming a viceroyalty in 1776, Argentina's colonial-era buildings were in general fairly modest structures, private houses or unadorned administrative buildings, rather than resplendent baroque palaces and cathedrals which could have plausibly formed the kernel of a grand national style— hence the need to draw on the heritage of its near neighbor, Peru. Martín Noel, *Contribución a la historia de la arquitectura Hispano-Americana* (Buenos Aires: Talleres S. A. Casa J. Peuser, 1921). For more on Noel, see Jorge Francisco Liernur, "Mestizaje, Criollismo, Estilo Propio, Estilo Americano, Estilo Colonial: Lecturas modernas de la arquitectura en América Latina durante el dominio español," in *Escritos de arquitectura del siglo XX en América Latina*, (Seville: Tanais, 2002), 79–111.

27 Roberto Schwartz, "Down with Tordesilhas!," *Portuguese Literary and Cultural Studies* 4/5 (2000): 277–293. For a culturalist perspective on the 1920s, see Patricia Funes, *Salvar la Nación: Intelectuales, cultura y política en los años veinte latinoamericanos* (Buenos Aires: Prometeo Libros, 2006).

28 For further discussion of this term see Jyoti Hosagrahar, *Indigenous Modernities: Negotiating Architecture and Urbanism* (London; New York: Routledge, 2005).

29 See Tenorio-Trillo, *Mexico at the World's Fairs*. On the 1929 Seville exhibition, see Eduardo Rodríguez Bernal, *Historia de la Exposición Ibero-Americana de Sevilla de 1929* (Seville: Ayuntamiento de Sevilla, 1994).

30 Jorge Schwartz, ed., *Brasil, 1920–1950: de la antropofagia a Brasília* (Valencia: IVAM Centre Julio González; Generalitat Valenciana, Conselleria de Cultura i Educación, 2000).

31 Gregori Warchavchik, "Apropos of Modern Architecture," in *Warchavchik e a introdução da nova arquitetura no Brasil: 1925 a 1940*, ed. Geraldo Ferraz (1925; São Paulo: Museu de Arte de São Paulo, 1965), English translation 264–265. Lucio Costa, "Razões da nova arquitetura," in *Lúcio Costa: Sôbre arquitetura*, ed. Alberto Xavier (1936; Porto Alegre: Centro dos Estudantes Universitários de Arquitetura, 1962), 17–41; Flávio de Carvalho, "A Cidade do Homem Nú," in *Flávio de Carvalho, arquitetura e expressionismo*, ed. Luiz Carlos Daher (1930; São Paulo: Projeto, 1982), 99–103.

32 Alfonso Pallares, Manuel Amábilis, Juan O'Gorman et al., *Pláticas sobre arquitectura* (1933; Mexico City: Instituto Nacional de Bellas Artes, 2001); Jose Villagrán Garcia, "Apuntes para un estudio, I: Objectivo de la Arquitectura," *Arquitectura* 3 (1939): 13–16; three additional articles in the series appeared in *Arquitectura* 4 (1940): 23–26; *Arquitectura* 6 (July 1940): 13–16; *Arquitectura* 8 (July 1941): 15–22.

33 Grupo Austral, "Grupo de arquitectos para el progreso de la arquitectura," *Nuestra Arquitectura* 6 (June 1939): 214–222; Agrupación Espacio, "Expresión de Principios de la 'Agrupación Espacio,'" *El Arquitecto Peruano* 11, no. 119 (June 1947): unpaginated.

34 Walter Gropius, ed., *Internationale Architektur*, Bauhausbücher I (Munich: A. Langen, 1925). The same Häring project was the only example of "Latin American" modernism to appear in Adolf Behne's *Der moderne Zweckbau* (1926). By contrast, Ludwig Hilberseimer's *Internationale neue Baukunst* (1927) and Bruno Taut's *Modern Architecture/Die neue Baukunst in Europa und Amerika* (1929) bypassed the region altogether.

35 Alberto Sartoris, *Gli elementi dell'architettura funzionale, sintesi panoramica dell'architettura moderna* (Milan: Ulrico Hoepli, 1932); the introduction to the second edition (1935) was written by Pietro Maria Bardi, who emigrated to Brazil with his wife, architect Lina Bo, in the mid-1940s.

36 Le Corbusier, *Precisions on the Present State of Architecture and City Planning*, trans. Edith Schreiber Aujame (1930; Cambridge: MIT Press, 1991).

37 See Christiane Crasemann Collins, "Urban Interchange in the Southern Cone: Le Corbusier (1929) and Werner Hegemann (1931) in Argentina," *Journal of the Society of Architectural Historians* 54, no. 2 (June 1995): 208–227.

38 Werner Hegemann, "Gemeinnützige Kleinwohnungsbauten in Buenos Aires," *Wasmuths Monatshefte für Baukunst und Städtebau* 16, no. 4 (1932): 185–192.

39 Hegemann, "Schinkelscher Geist in Südamerika," *Wasmuths* 16, no. 7 (1932): 333–341. The only architect mentioned by name is Mauricio Cravotto.

40 Hegemann noted the similarity of Berlin's *Laubenkolonien* to the "fringe settlements" of Buenos Aires—which, with their single-family houses and vegetable gardens, he considered to be preferable to tenements. Hegemann, "Als Städtebauer in Südamerika, 3: Der Seig der Randsiedlung über die Mietskaserne," *Wasmuths* 16, no. 5 (1932): 251, 250. Two additional articles appeared in issues 16, no. 3 (1932): 141–148, and 16, no. 4 (1932): 193–196.

41 Esther Born, "The New Architecture in Mexico," *Architectural Record* 81 (April 1937): 3. In the same year Born published an extended version of this material as a book under the same title.

42 F.A. Gutheim, "Buildings at the Fair," *Magazine of Arts* 32, no. 5 (May 1939): 316.

43 Philip L. Goodwin with photographs by G. E. Kidder Smith, *Brazil Builds: Architecture New and Old, 1652–1942* (New York: The Museum of Modern Art, 1943).

44 Sigfried Giedion, *A Decade of New Architecture* (Zürich: Girsberger, 1951), 3, emphasis in original.

45 As indicated by the list of CIAM affiliates in Giedion's book, Latin American membership closely tracked Sert's work in the region as an urban planner, with groups in smaller centers such as Peru and Colombia (Medellín as well as Bogotá), but not Mexico.

46 Alf Byden, "Report on Brazil," *Architectural Review* 108 (October 1950): 221, 222.

47 "Report on Brazil," *Architectural Review* 116 (October 1954): 237, 238, 239–240.

48 Bruno Zevi, *Storia dell'architettura moderna* (Turin: Einaudi, 1950), 285.

49 Henry-Russell Hitchcock, *Architecture: Nineteenth and Twentieth Centuries*, The Pelican History of Art (Baltimore: Penguin Books, 1958), 345–346.

50 See Patricio del Real, "Para caer en el olvido: Henry-Russell Hitchcock y la arquitectura latinoamericana," *Block* 8 (2011).

51 Leonardo Benevolo, *Storia dell'architettura moderna* (Bari: Laterza, 1960).

52 Sigfried Giedion, *Space, Time, and Architecture: The Growth of a New Tradition*, 4th ed. (Cambridge: Harvard University Press, 1962), xxxii.

53 Manfredo Tafuri and Francesco Dal Co, *Architettura Contemporanea* (Milan: Electa, 1976), 385.

54 Josep María Montaner, "La arquitectura moderna en Latinoamérica," in Leonardo Benevolo, *Historia de la arquitectura moderna* (Barcelona: Editorial Gustavo Gili, SA, 1982), 917–958.

55 See Kenneth Frampton, "Prospects for a Critical Regionalism," *Perspecta* 20 (1983): 147–162; and Frampton, "Towards a Critical Regionalism: Six Points for an Architecture of Resistance," in *The Anti-Aesthetic: Essays on Postmodern Culture*, ed. Hal Foster (Seattle: Bay Press, 1983), 16–30.

56 Kenneth Frampton, *Modern Architecture: A Critical History*, 2nd ed. (London: Thames and Hudson, 1985), 7.

57 Frampton, *Modern Architecture*, 4th ed. (2007), 377.

58 Ibid., 387.

59 Ibid., 389.

60 Valerie Fraser, *Building the New World: Studies in the Modern Architecture of Latin America, 1930–1960* (London; New York: Verso, 2000), 15.

61 Since then a number of similar institutions have been established: Instituto Carlos Arbeláez Camacho in Bogotá (1963); Instituto de Historia de Arquitectura in Montevideo (1985); Archivo de Arquitectura y Construcción de la Universidad de Puerto Rico in San Juan (1986); Centro de Información y Documentación Sergio Larraín García-Moreno in Santiago (1994).

62 Diego Angulo Iñiguez, *Historia del arte hispano-americano* (Barcelona: Salvat, 1945); *Ars Hispaniae; historia universal del arte hispánico* (Madrid: Editorial Plus-Ultra, 1947). In 1946, the US Society of Architectural Historians produced a special issue of its journal on Latin American architecture under the guidance of George Kubler; no modern architecture was included. *Journal of the Society of Architectural Historians* 5 (1945–1946).

63 Juan Pedro Posani, "Por una historia latinoamericana de la arquitectura moderna latinoamericana," *Boletín del Centro de Investigaciones Históricas y Estéticas* 9 (April 1968): 181–197. The participants were: Graziano Gasparini, Jesús M. Blanco, Victor M. Fossi, Erwin Walter Palm, Frederick Cooper Llosa, Fernando Chueca Goitía, Germán Téllez, Pedro Rojas, George Kubler, Leonardo Benevolo, Paolo Portoghesi, José García Bryce, Salvador Pinoncelly, Juan Pedro Posani, Sibyl Moholy-Nagy, and Ricardo de Robina. Absent from the conference were Francisco Bullrich, and Leopoldo Castedo, the Spanish architectural historian then exiled in Chile. In 1969, Castedo produced his wide-ranging survey *A History of Latin American Art and Architecture from Pre-Columbian Times to the Present*, Praeger World of Art Series (New York: Praeger, 1969).

64 Rafael López Rangel, *Arquitectura y subdesarrollo en América Latina* (Puebla: Universidad Autónoma de Puebla, 1975).

65 Roberto Segre, *América Latina en su arquitectura* (Paris: UNESCO, 1975); the book was subsequently published in English with Fernando Kusnetzoff as co-editor: Segre and Kusnetzoff, *Latin America in its Architecture*, trans. Edith Grossman (New York: Holmes & Meier, 1981).

66 Damián Bayón and Paolo Gasparini, *Panorámica de la arquitectura latino-americana* (Barcelona: Editorial Blume, 1977).

67 See Christian Fernández Cox's notion of an "appropriate modernity" which became the battle cry for an engaged historiography: "Hacia una modernidad apropiada: factores y desafíos internos," *Summa* 200–201 (September 1984), and "Modernidad apropiada en América Latina," *ARS* 11 (July 1989): 11–16. See also Marina Waisman's important contribution incorporating semiotics and typology into the discussion: "Algunos conceptos críticos para el estudio de la arquitectura latinoamericana," *Boletín del Centro de Investigaciones Históricas y Estéticas* 18 (April 1974): 153–160. For a critique of the historiography of the 1980s, see Adrián Gorelik, "Cien años de soledad? Identidad y modernidad en la cultura arquitectónica latinoamericana," *ARQ (Santiago)*, no. 15 (1990): 32–39. For an examination of the historiography from the 1980s onward, see Silvia Arango, "Historiografía latinoamericana reciente," *Archivos de Arquitectura Antillana: AAA* 9, no. 23 (January 2006): 163–169.

68 Ramón Gutiérrez, *Arquitectura y urbanismo en Iberoamérica* (Madrid: Ediciones Cátedra, 1983); Enrique Browne, *Otra arquitectura en América latina* (Mexico: G. Gili, 1988).

69 Yves Bruand, *Arquitetura contemporânea no Brasil* (São Paulo: Editora Perspectiva, 1981).

70 Marco Aurélio A de Filgueuras Gomes, ed., *Urbanismo na América do Sul: Circulação de idéias e constituição do campo, 1929–1960* (Salvador: Editora da Universidade Federal da Bahia, 2009). A Brazilian perspective on Latin America has also been explored by Ruth Verde Zein, "Siete motivos para la inexistencia de la crítica de arquitectura en América Latina, o aun la prueba de su existencia," *Summa+* 42 (2000): 123, and "Unidade Cultural: 4ª Bienal Ibero-Americana," *AU: Arquitetura e Urbanismo* 129 (2004): 26–31; as well as Carlos Eduardo Dias Comas, "Memorando latino-americano: La ejemplaridad arquitectónica de lo marginal," *2G. Revista Internacional de Arquitectura* 8 (1998): 10–20, and "20 casas latinoamericanas 1920–1970: Apuntes para una historia mal contada," *Equis* 1 (2008): 167–172.

71 Fernando Pérez Oyarzún, *Le Corbusier y Sudamérica: viajes y proyectos* (Santiago: Ediciones Arq, 1991). See also Ramón Gutiérrez, ed., *Le Corbusier en el Río de la Plata, 1929* (Buenos Aires: Centro de Documentación de Arquitectura Latinoamericana; Montevideo: Universidad de la República, 2009).

72 José Tavares Correia de Lira, *Warchavchik: fraturas da vanguardia* (São Paulo: Cosac Naify, 2011).

73 Alberto Sato Kotani, *José Miguel Galia, arquitecto* (Caracas: Universidad Central de Venezuela, 2002).

74 Enrique X. de Anda and Salvador Lizárraga Sánchez, eds., *Cultura arquitectónica de la modernidad mexicana: antología de textos 1922–1963* (Mexico City: Universidad Nacional Autónoma de México, Instituto de Investigaciones Estéticas, 2010); Abilio Guerra, *Textos fundamentais sobre história da arquitetura moderna brasileira*, 2 vols. (São Paulo: Romano Guerra Editora, 2010).

75 Jorge Francisco Liernur with Pablo Pschepiurca, *La red austral: obras y proyectos de Le Corbusier y sus discípulos en la Argentina (1924–1965)* (Buenos Aires: Universidad Nacional de Quilmes, 2008).

76 Lauro Cavalcanti, *As preocupações do belo: Arquitectura moderna brasileira dos anos 30/40* (Rio de Janeiro: Taurus Editora, 1995).

77 John A. Loomis, *Revolution of Forms: Cuba's Forgotten Art Schools* (New York: Princeton Architectural Press, 1999). A revised edition from 2011 reports on the impact that the book has had in promoting the preservation of the buildings.

78 Silvia Arango, *Historia de un itinerario* (Bogotá: Universidad Nacional de Colombia, 2002).

79 See Eduardo Luis Rodríguez and Gustavo Luis Moré, eds., "The Modern Movement in the Caribbean Islands," special issue, *Docomomo Journal* 33 (September 2005); Enrique Vivoni Farage, ed., *Klumb: Una arquitectura de impronta social/An Architecture of Social Concern* (San Juan: La Editorial, Universidad de Puerto Rico, 2006).

80 Luis E. Carranza, *Architecture as Revolution: Episodes in the History of Modern Mexico* (Austin: University of Texas Press, 2010). Patrice Elizabeth Olsen provides a significant contribution to the study of modern architecture in revolutionary Mexico from the perspective of a historian in *Artifacts of Revolution: Architecture, Society, and Politics in Mexico City, 1920–1940* (Lanham: Rowman & Littlefield Publishers, 2008).

81 Anahí Ballent, *Las huellas de la política: vivienda, ciudad, peronismo en Buenos Aires, 1943–1955* (Buenos Aires: Universidad Nacional de Quilmes, 2005).

82 Arturo Almandoz, *Urbanismo europeo en Caracas (1870–1940)* (Caracas: FUNDARTE; Ediciones de la Universidad Simón Bolívar, 1997); Adrián Gorelik, *La grilla y el parque: espacio público y cultura urbana en Buenos Aires, 1887–1936* (Buenos Aires: Universidad Nacional de Quilmes, 1998). In a similar vein, the transformative period of late nineteenth-century economic expansion and infrastructural transformation of the center of São Paulo is examined by Hugo Segawa in *Prelúdio da Metrópole: arquitetura e urbanismo em São Paulo na passagem do século XIX ao XX* (São Paulo: Ateliê Editorial, 2000).

83 Alberto Saldarriaga Roa, Alfonso Ortiz Crespo, and José Alexander Pinzón Rivera, *En busca de Thomas Reed: arquitectura y política en el siglo XIX* (Bogotá: Universidad Nacional de Colombia, 2005). Other recent efforts on nineteenth-century architecture

focus on the preservation of early industrial material culture, such as Cecilia Gutiérrez Arriola, *La revolución industrial y su patrimonio* (Mexico City: Universidad Nacional Autónoma de México, Instituto de Investigaciones Estéticas, 2007).

84 Robert González, *Designing Pan-America: US Architectural Visions for the Western Hemisphere* (Austin: University of Texas Press, 2011).

85 See Benedict Anderson, *Imagined Communities: Reflections on the Origin and Spread of Nationalism* (London; New York: Verso, 1983). For perspectives on Anderson's work from within the region, see Claudio Lomnitz, "Nationalism as a Practical System: Benedict Anderson's Theory of Nationalism from the Vantage Point of Spanish America," in *Deep Mexico, Silent Mexico: An Anthropology of Nationalism* (Minneapolis: University of Minnesota Press, 2001), 3–34; John Charles Chasteen and Sara Castro-Klarén, eds., *Beyond Imagined Communities: Reading and Writing the Nation in Nineteenth-Century Latin America* (Washington DC: Woodrow Wilson Center Press; Baltimore: Johns Hopkins University Press, 2003).

Part I

SINGULAR JOURNEYS

1

LUCIO COSTA'S LUSO-BRAZILIAN ROUTES

Recalibrating "Center" and "Periphery"

Gaia Piccarolo

Rhetoric concerned with identity-building and the relationship with European culture in terms of "periphery" versus "center" is one of the leitmotivs of Latin American architectural historiography. Facing the demand to mediate between local, national, and international levels, categories like regionalism, nationalism, and Pan-Americanism entered the architectural debate from the first decades of the twentieth century, in response to the search for a recognizable identity.

Lucio Costa, a key protagonist of these debates, worked strenuously in order to better define Brazilian particularities.[1] While he is currently portrayed as striving for a synthesis between modernity and tradition, international and national architectural languages, less known is his contribution to the forging of relationships and exchanges between Brazil and its former colonizer, which deserves further investigation on the basis of its economic, political, and diplomatic dynamics, and as an example of the circulation of ideas and models across the Atlantic. This chapter explores the reciprocal influences between Brazil and Portugal across their routes to modernism, especially through Costa's personal contribution to the development of Luso-Brazilian relationships. In the process, the dynamics of "center" and "periphery" are brought into question, pointing out the specificity of this case study due to its peculiar historical developments and motivations.

Costa undertook his first study trip to Portugal in 1948—from a supposed peripheral condition in Brazil—in order to understand the process through which Brazilian tradition had developed from the center of its source.[2] Yet Portugal was already looking at Brazil as a new center for architectural innovation: five years earlier, the Museum of Modern Art, New York, had held its seminal exhibition on Brazilian architecture, whose accompanying catalogue, *Brazil Builds* circulated in Portugal, as in the rest of Europe, spreading expressive images of Brazilian architectural production.[3] It is worth noting that this process inverted that of the first decades of the twentieth century, when the Portuguese legacy was a main topic of discussion during Brazil's passionate "anthropophagic" search for its own cultural roots, led by poet Oswald de Andrade, and when the migration of Portuguese intellectuals to Brazil continued to bring to the former colony ideas and issues from contemporary Portuguese debates.

This was the case with the engineer Ricardo Severo,[4] the main person responsible for the recovery of the Portuguese legacy in Brazil and for the importation of the debate on the "Portuguese house," led in those years by the traditionalist Portuguese architect Raul Lino, whose articles circulated in Brazilian architectural periodicals during the 1920s.[5] The young Costa played an active role in the traditionalist movement in Rio de Janeiro, mainly through his personal and professional contacts with its leading figure José Mariano Filho,[6] whose research was focused on the formulation of a model of the "Brazilian house" inspired by domestic architecture of the colonial period. This effort was part of the search for an "authentic" and "national" architectural expression that derived from a widespread discomfort with nineteenth-century eclecticism and French-based academicism.

In 1924, within the framework of a renewed and wider interest in national historical heritage,[7] Mariano Filho sponsored Costa to undertake a trip to Diamantina, a colonial city in the valley of Minas Gerais, in order to study its architectural heritage; on Costa's return he observed in an interview how he had been deeply touched by the contact with vernacular colonial domestic architecture.[8] The drawings produced show a careful and technical-focused survey of authentic traditional architectural and constructive solutions, intended for collecting a sort of repertoire for use in current design practice.[9] After all, Costa's architecture in the 1920s was perfectly placed within the diffuse traditionalism whose circulating models nationalized Beaux-Arts composition through the incorporation of colonial decorative motifs.

Some years after this first mission, on returning from a study trip to Europe (1926–1927), Costa was in Minas again, visiting Caraça, Sabará, Mariana, Ouro Preto, and going back to Diamantina. This time free from any specific commission, he immortalized several monumental buildings in their historical sites, fixing on paper architectural details, furniture pieces, even landscapes with passionate accuracy and a charming pencil stroke.[10]

At this precise moment, the sociologist Gilberto Freyre, one of the main theoreticians of Portuguese influence on Brazilian identity in the complexity of its cultural manifestations, was promoting the ideal of a "national unity under regional variety,"[11] a theme that also motivated Costa's 1929 text: "O Aleijadinho e a arquitetura tradicional." Despite his skepticism concerning the refined and decorative expression of the artist from Minas, Costa's essay asserted that upon examining the national architectural heritage, one comes to realize that "Brazil, despite its extension, its local differences, and other complications, had really to be one thing. Good or bad, it has been shaped all-in-one, by the same spirit and by just one hand."[12]

This mix between nationalism and the emphasis on regional specificities was one of the many points where the aims of intellectuals and the government converged at this moment in Brazil. A revolution in 1930 established the first regime of Getúlio Vargas (1930–1945, 1951–1954), marked by a strong centralization of the state, an aggressive nationalism, and a push toward modernization. Within the soft renovation of cultural institutions promoted by the new government, modernist intellectuals who were engaged with the Vargas administration called upon Costa to lead the School of Fine Arts, a post that he held for less than one year, and that inaugurated his modernist militancy in relation to local artistic and architectural debates. Over the following years he became deeply involved in the dissemination of modernism and of the ideas of European modernist architects— a process culminating with his coordination of the project for the Ministry of Education and Health in Rio de Janeiro.

Costa's new distance from the ideological and aesthetic positions of the traditionalists became evident in a meeting with Portuguese architect Raul Lino, who, when visiting Brazil in 1935, insisted on being introduced to the man whom he characterized as a "true mentor of the young Brazilian architects."[13] In Lino's retrospective reconstruction of their dialogue, mainly afforded through quotations

1.01 Lucio Costa, drawing of Aleijadinho's Igreja do Carmo, Ouro Preto, Brazil, 1927

from Costa's text "Razões da nova arquitetura,"[14] he affirmed—partially mystifying the truth yet stressing well the overwhelming distance between their positions—that his "distinguished colleague . . . doesn't want to hear about tradition,"[15] but rather approached problems from a strictly Functionalist point of view.

Meanwhile, the first Vargas government was marked by efforts at rapprochement with Portugal, further emphasized after Vargas's authoritarian turn in 1937 with the institutionalization of the Estado Novo (New State), modeled on the Salazar regime in Portugal (1932–1968).[16] The creation of the Serviço do Patrimônio Histórico e Artístico Nacional (SPHAN) in the same year, represented an important instrument of the government's cultural strategy to establish stronger relationships with the former colonizers within the framework of Brazil's affirmation as a world power.

Under the sponsorship of the Ministry of Education and of SPHAN, several modernist intellectuals were involved in building a theoretical definition of a

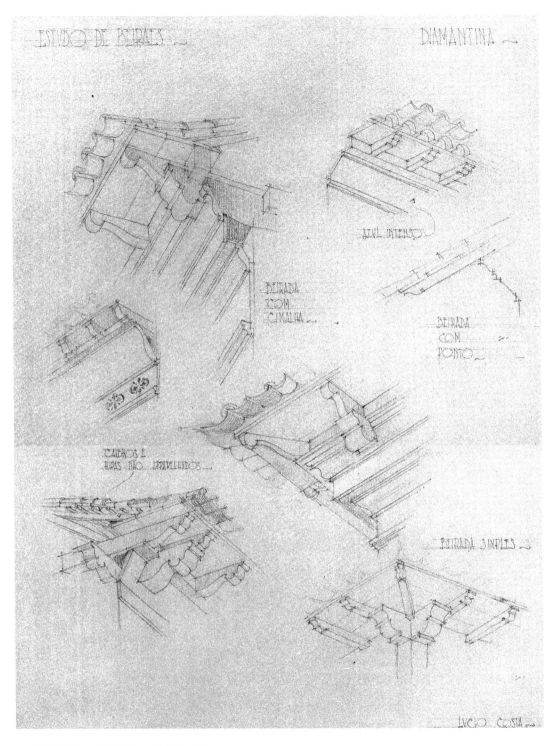

1.02 Lucio Costa, studies of eaves in Diamantina, 1924

"national tradition" on which the selection of the historical sites to be physically safeguarded would be grounded. Costa's involvement with this organization was a turning point in determining the direction of his interests in the following years, resulting in the definite development of a personal modern syntax that synthesized a local and vernacular vocabulary.[17]

1.03 (above) Lucio Costa at the Serviço do Patrimônio Histórico e Artístico Nacional (SPHAN)

1.04 Lucio Costa, Park Hotel São Clemente, Nova Friburgo, Brazil, ca. 1948

His article "Documentação necessária," published in 1937 in the first issue of SPHAN's official journal, returned to the theme of the colonial house.[18] However, the protagonist was now its direct progenitor, the architecture of the "good Portuguese tradition," together with the "old Portuguese master builder of 1910" who, in Costa's opinion, was able to instill in his works a "perfect plastic health" and a "typical Portuguese *carrure* [build or stature]." Costa also mentioned the importance of studying the vernacular Portuguese house in order to "take advantage of the lessons of its more than three hundred years of experience," which, in one of the illustrations, he defined as "pure Le Corbusier" due to the modernity of its architectural solutions.

Within SPHAN, Costa became involved in day-to-day work on the historical heritage of Brazil, starting with a field study of the ruins of Sete Povos das Missões in Rio Grande do Sul, the result of which was the little museum he designed in order to preserve the site's archeological remains. Although its horizontal tile roof and open arcades—whose columns incorporated capitals and materials found in situ—were clearly inspired by old colonial houses, the steel-and-glass Miesian box containing the archaeological finds allowed them to be perceived as part of the landscape and gave a decisively modern tone to the whole intervention.

At the same time, he published articles addressing questions such as the evolution of Luso-Brazilian furniture and the architecture of the Jesuits in Brazil, as well as several historical essays on Aleijadinho, in which he retreated from his earlier skepticism.[19]

Considering the body of Costa's written work in this period and his untiring promotion of Oscar Niemeyer, it is easy to discern a theoretical formula aimed at creating an ideal bridge between different and distant experiences in order to prove the existence of a unique "national personality."[20] On one side, the unselfconscious constructive rationality of vernacular local architecture and its sober aesthetics, free from rigid compositional schemes, were associated with the modernist principle of correspondence between form and function. On the other side, the personalities of Aleijadinho and Niemeyer, poets par excellence of the baroque voluptuous curve, were identified as select representatives of the "native artistic genius."[21]

Three important events at the turn of the 1940s highlighted the distance between the architectural panorama of Brazil and Portugal: the Exhibition of the Estado Novo held in Rio de Janeiro in 1938, the New York World's Fair of 1939, and the Exhibition of the Portuguese World held in Lisbon in 1940, which celebrated the dual anniversaries of the foundation of Portugal as an independent kingdom (1140) and the Restoration of Independence from Spain (1640).

The exhibitions of 1938 and 1939 can be considered as different steps in the gradual identification of the Brazilian modernist language as representative of state authority, which would be consecrated in the following decade with Pampulha and would reach its highest moment with Brasília. At the Exhibition of the Estado Novo the model of the Ministry of Education and Health was exhibited side by side with other governmental projects of the period, whose heterogeneous architectural languages adhered to academic and neocolonial models.

The Brazilian pavilion at the New York World's Fair, designed by Costa and Niemeyer, was a strictly Corbusian horizontal building on *pilotis*.[22] However, the use of a distinctive sun-screening system—a synthesis between a *brise-soleil* and a Brazilian *cobogo* (perforated ceramic bricks which are an adaptation of the Arab *muxarabi*)—the curvilinear flow of one of the facades, the mezzanine and the ramp, and the treatment of the external area as a tropical garden, anticipated the free sensuality of Pampulha. Together with those of Finland and Sweden, this pavilion was among those most appreciated by the international critics, seen as embodying an architectural expression at once "modern" and "national," where the rigidity of modernist orthodoxy was avoided without explicit references to the past.

1.07 Model of the Ministry of Education and Health at the Exhibition of the Estado Novo, Rio de Janeiro, 1938

1.08 Raul Lino, Brazilian Pavilion at the Exhibition of the Portuguese World, Lisbon, 1940. Photograph by Mário Novais

At the Exhibition of the Portuguese World, Brazil was represented by a pavilion designed by Portuguese architect Raul Lino: its colonnade—resembling the stylized trunks of enormous palms with light filtered through the ceiling—evoked the exotic atmosphere of a tropical forest. Although it could be considered one of Lino's less traditional works, it had traces of the same abstract monumentality that characterized the other pavilions, described by Ana Tostões as the triumph of a "historical and regional vocabulary applied to a classical-based architectural language."[23] The exhibition marked the end of an early modernist phase in Portugal, spanning the mid-1920s to the end of the 1930s, when the Salazar regime co-opted the new language in order to impel a symbolic renewal via official architecture.

Considering the cultural isolation of Portugal under Salazar, which contributed to delaying the country's adoption of modernism, it is clear why Brazilian modern architecture played such a pivotal role in the country's architectural renewal, evincing the inversion of "center" and "periphery." A few years after the Exhibition of the Portuguese World, the catalogue of *Brazil Builds* circulated in Portugal, where the innovative experiments undertaken in the former colony were welcomed with enthusiasm by a young generation of professionals eager to overcome the limitations imposed by the regime's architectural program, dominated at that moment by the recuperation of traditionalist idioms and celebratory monumentality.[24]

In this reception of Brazilian architecture, Costa emerged—despite his limited number of executed modern buildings—as a sort of "spiritual guide" of the architects who had driven the transformations, and his ideas circulated through architectural networks via the publication of some of his essays in Portugal.[25] The search for an architectural expression that was both modern and national found a triumphant example in the work of these architects, and Costa was a key theoretician of a reading of vernacular tradition that avoided the limitations of historicism.

The dichotomy of tradition and modernity was also evident in the stance taken by Philip Goodwin and G. E. Kidder Smith in *Brazil Builds*: in presenting the colonial past and contemporary modernism, praising an aesthetics of sobriety and horizontal lines while criticizing nineteenth-century eclecticism, they were clearly supporting Costa's preferences.[26]

The circulation of these ideas had a direct effect in Portugal: in 1947, the Portuguese architect Francisco Keil do Amaral published the article "Uma iniciativa necessária," echoing Costa's 1937 text in both title and intentions, without directly referring to it.[27] Keil do Amaral encouraged Portuguese architects to engage in a deep study of "our regional architecture," in order to capture its clear functionality and intrinsic ability to adapt to the environment, instead of reproducing its decorative elements. A later consequence of this call to action was the Inquérito à Arquitectura Regional Portuguesa (1955–1960), a campaign to study and survey vernacular architecture that was carried out by several groups of young Portuguese architects.

Meanwhile, Costa put into practice his own project of studying the architectural heritage of the Portuguese interior: in August 1948, the director of SPHAN requested from the Minister of Education, "that architect Lúcio Costa . . . be designated to go to Portugal, to proceed with studies that have turned out to be necessary in order to elucidate the capital points of Portuguese influence in the formation and evolution of the plastic arts in Brazil."[28] In September, Costa embarked for Lisbon, where he stayed four months. This was certainly not an isolated endeavor, since in December 1948 the Brazilian Foreign Minister Raul Fernandes signed, during his visit to Lisbon, an agreement of intellectual cooperation for the promotion of "a large interchange between educational institutions and men of culture of both countries."[29]

In an unfinished written report, *Introdução à um relatório* (published only in 1995), Costa presented the results of this first mission to Portugal but also declared the failure of its initial purpose: his observation of the complexity characterizing the evolution of architectural typologies in the different regions of Portugal suggested that it was impossible to establish coherent lines of derivation for their colonial developments.[30] This observation allowed Costa to conceive of an independent development of Portuguese architecture in the colony—one that reproduced the same "capricious interferences" characterizing the earlier evolutions of Portuguese architecture in its homeland, but demonstrated its "own personality" and was as authentic and legitimate as the original. This idea was further developed in Costa's contribution to the Luso-Brazilian Congress in Washington DC in October 1950, in which he suggested interpreting the architectural masterpieces of Minas Gerais not only as a manifestation of an autonomous elaboration of Portuguese models, but also as "the more original and worthy contribution for the production of Portuguese art in the final phase of the baroque cycle."[31]

At the beginning of 1952—the same year in which Gilberto Freyre traveled through the Portuguese provinces in order to look for what he defined as the "Portuguese constants of character and action"[32]—Costa undertook a second trip to Portugal. The aim of this six-month mission was once again to carry out "studies on the relationships between plastic arts in that country and in Brazil, as a complement to the notable works he had already carried out in this respect."[33] However, the intensification of Costa's international commitments (mostly related to UNESCO)[34] caused several delays in the fulfillment of the official goals of the mission, which as a result was extended until at least January 1953.[35] From several letters to Rodrigo Mello Franco de Andrade it is possible to gather—in addition to detailed reports and intriguing considerations on some of his visits and meetings— an emerging discouragement, due to the continuous displacements and the difficulties of carrying out research in a not always hospitable country.[36] The architect's sketchbooks from the trip provide precious information about the itinerary, the subjects of the inquiry and the methodological approach: a multitude of little rapid-stroke sketches, devoid of any artistic complacency, demonstrate a perceptive faculty in detecting local typologies, constructive and decorative elements; the more realistic effect of the academic chiaroscuro gives way to the synthetic and stenographic outline of the architectural abstract volumes, achieved through few expressive signs and completed by several textual annotations.

It is worth considering the prospective advantages of a funded visit to Europe for the consolidation and the expansion of Costa's international network:[37] this process had begun with Le Corbusier's visit to Rio de Janeiro in 1936, Giedion's invitation to Costa to become an official member of CIAM in 1937, and the New York experience of 1938–1939. While Costa's obligations in Italy and France gave him the opportunity to establish professional relationships outside Brazil, there is little evidence of any direct contacts with Portuguese architects, or of any official events he may have attended during these first two visits in 1948 and 1952.[38] Certainly he was not directly involved in the main events that marked the popular dissemination of Brazilian architecture in Portugal, which he missed by just a few months: the first exhibition dedicated to Brazilian architecture, organized in Lisbon in February 1949 by the Brazilian architect Wladimir Alves de Souza, and the UIA Congress held in Lisbon in September 1953, which included among other events an exhibition of contemporary Brazilian architecture. The latter, however, gave significant space to Costa's activities both as architect and as promoter of modern architecture.[39]

1.09 Lucio Costa, sketches from the study trip in Portugal, 1952

1.10 Lucio Costa, sketches
from the study trip in
Portugal, 1952

The existence of a network of professional and personal relationships between Costa and Portuguese architects only becomes evident in the early 1960s, when the first traces of Costa's correspondence with Carlos Ramos are found. The turning point was without doubt the inauguration of Brasília: the new capital immediately became a regular destination on the study trips of young Portuguese architects, clearly determining an influence on their architectural production. Moreover, Costa had met Ramos, one of the protagonists of debates over the renewal of architecture in Portugal, during the first of his many trips to Brazil undertaken in order to develop the project for the Portuguese embassy in Brasília. This meeting happened in 1961, when this important charge pushed Ramos to visit the city's planner in his office in the Ministry of Education in Rio de Janeiro.[40] However meeting Ramos's project, later appreciated by Costa, was never realized.[41]

That same year, Ramos invited Costa to coordinate a two-week workshop at the Porto School of Fine Arts, whose consequences for the development of Portuguese architecture still need to be investigated.[42] On the occasion of this visit, Costa and his daughter Helena were photographed visiting the market of Barcelos, and, together with Ramos, José Carlos Loureiro and a young Fernando Távora, leaving from the Porto station of São Bento.[43] An article by Antonio Menéres, a pupil of Ramos, elucidates the role that Ramos played as mentor to the new generation of architects within the Porto school, disseminating Costa's ideas on the importance of plastic expression in architecture and on the synthesis of the arts.[44] The article also mentions a visit to Trás-os-Montes during which Lucio and Helena Costa were accompanied by a group of Portuguese architects who had been involved in the Inquérito of the late 1950s.[45]

The results of the Inquérito, published in 1961 as *Arquitectura popular em Portugal*, rejected the idea of identifying a unique national style, which Raul Lino's concept of the "Portuguese house" had claimed to embody.[46] The attitude of looking at the lines of influence in their complexity, trying to go back to the formation of the different regional iterations of vernacular architecture throughout Portugal, has much in common with Costa's conclusions from his first trip in 1948. It is not surprising that Costa later showed a full understanding of the direct influence of his positions upon this endeavor: a commentary introducing "Documentação necessária" in his autobiographical volume *Registro de uma vivência*, stressed the fact that Ramos had affirmed that the Inquérito was a direct consequence of the ideas Costa had put forward in 1937.[47] On the other hand, the reciprocal influence becomes evident in the third edition of *Arquitectura popular em Portugal* (1988), in which Nuno Teotónio Pereira explicitly refers to Brazilian architects "who are interested in the Portuguese roots of their architecture," and points out that they considered the book—as the Portuguese did with *Brazil Builds* in the 1950s—"a rare treasure."[48]

It seems paradoxical that the only two projects which we can affirm with certainty Costa designed for Portugal[49]—the social housing complex for the Quinta do

1.11 Lucio Costa, study for the monument to Infante Dom Henrique (Henry the Navigator), Sagres, Portugal, ca. 1954

Rouxinol (Seixal) commissioned by Lucio Thomé Feteira in 1953, and the monument to Henry the Navigator (1394–1460) in Sagres, conceived for the 1954 international competition[50]—both remained on paper, although they were subsequently proposed in revised form as projects for the new Brazilian capital. The Quinta do Rouxinol's lozenge-shaped layout was adopted in the proposal for the Quadras Econômicas in Brasília, while the monument—which in some ways condensed Costa's research on monumental expression—contained in embryonic form many of his subsequent monumental projects, including Brasília's television tower. Dedicated to one of the heroes of the Portguese Age of Discovery, the monument to Henry the Navigator was a highly nationalistic project whose main aim was to express a symbolic commitment to the memory of the glorious period of the Portuguese Empire, achieved through the use of an explicitly modern language. In the draft of his project proposal, Costa explained that although he

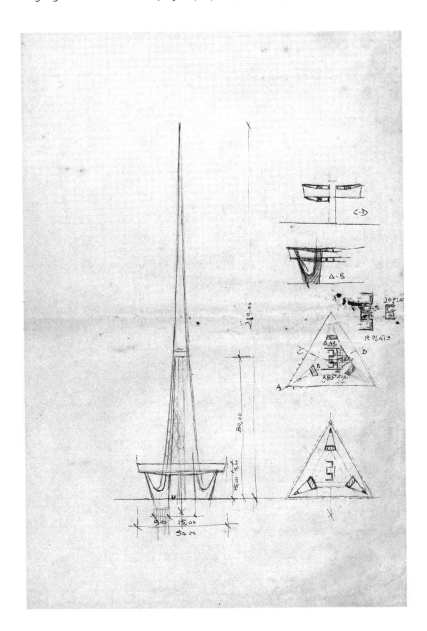

1.12 Lucio Costa, studies for the television tower, Brasília, 1957–1959

was informed quite late about the competition, he decided to submit his proposal, seeing it as the only appropriate solution for the site: in his words, the memory of his visit to Sagres and the careful observation of the photographs provided had allowed him "to *feel* the problem and to clearly *see* the solution."[51] Stressing the need to adopt "simple and pure forms, as much as possible detached from the ground in order to be discerned as a plastic gesture clearly perceivable in the distance, whose noble intention is to spiritualize the bare and rude beauty of the landscape," Costa conceived a gigantic, more or less trapezoidal-based pyramid emerging from a triangular concrete platform suspended on three pillars. It should be noted that, despite the difference in scale and function, Brasília's television tower shows a direct derivation from this scheme; however, a more emphatic verticality allows the television tower to work as a landmark contrasting the horizontal development of the city's monumental axis, a panoramic view of which can be enjoyed from the observatory on the tower's platform.

In the case of Sagres, the verticality of the whole and the detachment of the platform from the ground—even though the crypt signified the monument's material and spiritual grounding in that particular place—symbolized the main vocation of Henry the Navigator and evoked the historical exchanges and the ideal connection with the other side of the Atlantic. The interplay of different materials—steel and glass, granite from northern Portugal, local limestone, stone from the region of Lisbon, sheets of crystal painted with lead—and the changing relation with the site at the different levels—almost a ritual path of re-appropriation of the site—aimed to articulate the genius loci proper to the place, as Costa's direct reference to the "light of Extremadura" testifies. The draft also mentioned the will to express the "feeling of the Portuguese people," which was, significantly, linked to the "feeling of this ancient colony," stressing the profound relationship between the two cultures and integrating the monument into the ongoing construction of a Brazilian artistic and cultural identity.

It is interesting to observe that the synthetic curvilinear gestures of the winning project by João Andersen[52]—more intrusive than Costa's proposal in relation to the site—curiously reveals a certain correspondence with the simplified monumentality that Niemeyer was to develop in Brasília a few years later.

As a conclusion to these observations, we may note that although Costa's desire to outline an authentically Brazilian artistic character through the identification of parental relationships with the former colonizers did not produce the expected results, the cultural exchange between the two countries bore fruit in different and unforeseen ways. Both Brazil and Portugal—at different stages of their processes of modernization and architectural renewal—looked to each other for answers. Costa looked for the roots of his culture at the presumed center of its formation in order to give depth to his search for a national architectural expression; he looked at the past, a past that remained silent or even suggested more enigmas rather than revealing the truths that he had expected to discover. On the other side, the Portuguese looked at Costa and at Brazilian free-form modernism as the future they wanted to achieve. The parallel political situations of the two countries at this time prepared the ground for a new, pragmatic relationship, and the migration of people and of ideas achieved the rest. What had been the "periphery" turned out to be one of the "centers" from which architectural modernism spread out with new energy.

The effect of this circulating migration was two-sided in Portugal, which had the ambiguous position of being the former colonizer of a new world power, and at the same time a peripheral figure in European cultural debates. For this reason its process of assimilating avant-garde architectural movements was analogous

1.13 Lucio Costa, study for the monument to Infante Dom Henrique (Henry the Navigator), Sagres, Portugal, ca. 1954

1.14 João Andersen, Mar
Novo, project for the Sagres
monument, 1954

to that of Brazil in the early 1920s, although their subsequent history differed in
line with the specificities of each context.

In any case, the migration of architectural ideas from the "former-periphery" to
the "former-center" that began with *Brazil Builds* was not the only direction of
exchanges between the two sites. At least from the early 1960s, when on one
side Brasília marked both the peak and the exhaustion of the heroic phase of
Brazilian modernism, and, on the other, the Inquérito bore its fruit, the intellectual
circulation worked in both directions.

In this two-way exchange, the example of Lucio Costa is highly significant. His
theoretical positions as well as his architectural practice, filtered through the
experience of the Inquérito and the specific Portuguese route to modernity,
have had a strong influence on subsequent developments in some quarters of
Portuguese architecture; for example, in the work of Fernando Távora.[53] Costa
and Távora share, for instance, an interest in the historical and contextual roots
of architecture—in the new and in intervention in the old—as well as the principle
of a "third way" mediating between the abstraction of modernism and a
traditionalist historicism, between high and popular culture.

On the other hand, it is certainly reasonable to assume that Costa himself had
feedback from his Portuguese reconnaissance missions, resulting from both his
prior expectations and his day-to-day experiences in situ, including direct contacts
with local architects. On a scientific level, he did not manage to gather his
reflections into a systematic work, nor to build a systematic method of analysis,
and his argument remained in an unfinished state. In any case, his direct and
recurring contacts with Portugal, considered in continuity with his earlier
experiences of colonial Brazil, gave new substance and inspiration to the synthesis
between modernity and tradition that he tried to pursue from the end of the
1930s, almost twenty years ahead of Portugal.

The house he realized in Rio de Janeiro in 1980 for his daughter Helena stands as
one example of this continuity. At first sight it is characterized by a classical

facade suggesting both Palladian and modernist influences; its balconies are suspended within an orthogonal diaphragm recalling some Italian Rationalist masterpieces. With a more in-depth examination, its clear white volumes and generous glass partitions create a harmonious dialogue with the traditional tile roof and the interior decorations in ceramic tile and hand-crafted wood; the fluctuating, white space of the house is filtered, but never interrupted, by delicate walls of *treliças* (latticework), and tropical vegetation invades the private inner patio. Costa described it as "a sincerely contemporary house, albeit with a look and *saudade* [nostalgia] of our past. A *Brazilian house*: something the neocolonial was not able to convey."[54]

Notes

1   Born in Toulon in 1902, Costa was personally situated in between the old and the new world. Educated in England and Switzerland, he moved to Rio de Janeiro in 1916, where he studied at the School of Fine Arts.

2   However, Costa had already visited Lisbon in 1926, during his post-graduation tour of Europe.

3   Philip Goodwin and G. E. Kidder Smith, *Brazil Builds: Architecture Old and New 1652–1942* (New York: Museum of Modern Art, 1943). For an in-depth examination of the *Brazil Builds* exhibition and its consequences, see Zilah Quezado Deckker, *Brazil Built: The Architecture of the Modern Movement in Brazil* (London; New York: Spon Press, 2001). According to Ana Vaz Milheiro, Nuno Teotónio Pereira introduced the book into Portugal around 1945. See Milheiro, *A Construção do Brasil. Relações com a cultura arquitectónica portuguesa* (Porto: FAUP, 2005).

4   See Joana Mello, *Ricardo Severo: da arqueologia portuguesa à arquitetura brasileira* (São Paulo: Annablume Editora, 2007).

5   Raul Lino, "Como projectar uma casa de moradia," *A Casa* 7 (November 1924): 10–11.

6   José Mariano Filho (1881–1946) played a major role in the cause of national architecture, being active within several artistic institutions, such as the Sociedade Nacional de Belas Artes and the Instituto Brasileiro de Arquitetos. See Augusto da Silva Telles, "Neocolonial: la polémica de José Mariano," in *Arquitectura neocolonial: América Latina, Caribe, Estados Unidos*, ed. Aracy Amaral (São Paulo: Fundação Memorial da América Latina, 1994), 237–248.

7   Several modernist artists and intellectuals, such as Mário de Andrade, Oswald de Andrade, and Tarsila de Amaral, traveled across Minas Gerais in the same years in which Costa and other young architects from Rio de Janeiro and São Paulo were undertaking their study missions.

8   "Um architecto de sentimento nacional," *A Noite*, June 18, 1924.

9   These drawings are housed in the Museum Casa de Juscelino in Diamantina.

10 While in Minas, Costa also used this time to work on the competition for the Argentine embassy in Rio de Janeiro. These drawings are kept in the Casa de Lucio Costa (hereafter CLC).

11 Gilberto Freyre, *Manifesto regionalista de 1926* (Recife: Instituto Joaquim Nabuco de Pesquisas Sociais, 1967), 33.

12 Lucio Costa, "O Aleijadinho e a arquitetura tradicional," in *Lúcio Costa: Sôbre arquitetura*, ed. Alberto Xavier (Porto Alegre: Centro dos Estudantes Universitários de Arquitetura, 1962), 15–16. Note that Freyre quoted Costa's sentimental evocations of the colonial house in this text in the introduction of the first edition of *Casa Grande e Senzala* (1933).

13 Raul Lino, *Auriverde jornada. Recordações de uma viagem ao Brasil* (Lisbon: Edição de Valentim de Carvalho, 1937), 91.

14 Lucio Costa, "Razões da nova arquitetura," in *Lúcio Costa: Sôbre arquitetura*, 17–41, first published in *Revista da Diretoria de Engenharia da Prefeitura do Distrito Federal* 1, vol. III (January 1936).

15 Lino, *Auriverde Jornada*, 91–95.

16 José Calvet de Magalhães, *Breve história das relações diplomáticas entre Brasil e Portugal* (São Paulo: Paz e Terra, 1999), 84. The first Congress of Portuguese people in Brazil took place in 1931, and in 1933 the two countries signed the first trade treaty after Brazil's independence.

17   Costa worked in SPHAN from the beginning of its activity in 1937. In 1946 he became chief of the Divisão de Estudos e Tombamentos, a post he held until 1972.

18   Lucio Costa, "Documentação necessária," *Revista do SPHAN* 1 (1937): 31–39. For an English translation see Costa, "Necessary Documentation," *Future Anterior* 6, no. 2 (Winter 2009): 48–57.

19   Lucio Costa, "Notas sobre a evolução do mobiliário Luso-Brasileiro," *Revista do SPHAN* 3 (1939): 149–162; Costa, "Arquitetura Jesuítica no Brasil," *Revista do SPHAN* 5 (1941): 9–100; and Costa, *Registro de uma vivência* (São Paulo: Empresa das Artes, 1995), 520–552.

20   Lucio Costa, "Imprévu et importance de la contribution des architectes brésiliens au développement actuel de l'architecture contemporaine," *L'Architecture d'aujourd'hui* 42–43 (August 1952): 4–7.

21   Ibid.

22   On Brazil–US relations see Lauro Cavalcanti, "Architecture, Urbanism, and the Good Neighbor Policy: Brazil and the United States," in *Latin American Architecture 1929–1960: Contemporary Reflections*, ed. Carlos Brillembourg (New York: Monacelli Press, 2004), 50–59.

23   Ana Cristina Tostões, "Moderno e nacional na arquitetura portuguesa. A descoberta da modernidade brasileira," in *Moderno e Nacional*, ed. José Pessoa et al. (Niterói: EdUFF, 2006). The exhibition also included a "Portuguese village" which reproduced local vernacular architecture.

24   See Tânia Beisi Ramos and Madalena Cunha Matos, "Recepção da Arquitectura Moderna Brasileira em Portugal—registos e uma leitura," in *Moderno e Nacional*, 164–166.

25   Ibid.

26   Costa, being among those met by Goodwin and the photographer Kidder Smith on their Brazilian trip, later expressed his surprise to see what he considered his personal position assumed as an original statement in the catalogue of the exhibition. Lauro Cavalcanti, *Moderno e Brasileiro. A história de uma nova linguagem na arquitetura (1930–1960)* (Rio de Janeiro: Jorge Zahar Editor, 2006).

27   Francisco Keil do Amaral, "Uma iniciativa necessária," *Arquitectura* 14 (April 1947): 12–13.

28   Rodrigo Mello Franco de Andrade to Clemente Mariani Bittencourt, August 12, 1948, Coleção Personalidades, Série Lucio Costa, Arquivo Noronha Santos, IPHAN (hereafter ANS/IPHAN).

29   Calvet de Magalhães, *Breve história*, 88.

30   Costa, *Registro de uma vivência*, 456.

31   Lucio Costa, "A Arquitetura Brasileira Colonial," in *Proceedings of the International Colloquium on Luso-Brazilian Studies* (Nashville: The Vanderbilt University Press, 1953), 121–122.

32   Gilberto Freyre, *Aventura e Rotina: sugestões de uma viagem a procura das constantes portuguêsas de caráter e ação* (Rio de Janeiro: José Olympio, 1953).

33   Rodrigo Mello Franco de Andrade to Clemente Mariani Bittencourt, December 13, 1951, ANS/IPHAN.

34   In particular, the International Conference of Artists promoted by UNESCO in Venice on September 1952 and the consultancy for the UNESCO Headquarters in Paris (with Le Corbusier, Walter Gropius, Sven Markelius, and Ernesto Nathan Rogers). At the same time Costa was committed to the project for the Brazilian student residence for the Cité Universitaire in Paris, later developed by Le Corbusier.

35   Lucio Costa to the Portuguese Consulate in Lisbon, January 7, 1953, ANS/IPHAN.

36   The letters and sketchbooks from the trip are kept in CLC.

37   It seems that his presence in Venice in 1952 depended on the founding of his mission to Portugal: Rodrigo Mello Franco de Andrade to Paulo E. de Berrêdo Carneiro, November 28, 1951, ANS/IPHAN.

38   In 1950, Costa received honorary membership in the National Syndicate of Architects of Lisbon (ANS/IPHAN).

39   Ramos and Matos, "Recepção da Arquitectura Moderna Brasileira."

40   Carlos Ramos to Lucio Costa, Lisbon, February 15, 1962, CLC.

41   Tânia Beisi Ramos, Madalena Cunha Matos, *Um encontro, um desencontro. Lúcio Costa, Raul Lino e Carlos Ramos*, http://www.docomomo.org.br/seminario%207%20pdfs/034.pdf (Proceedings of the 7º Docomomo Brasil, Porto Alegre 2007).

42 See also Octávio Lixa Filgueiras to Lucio Costa, Porto, February 19, 1965, CLC.

43 *Carlos Ramos, exposição retrospectiva da sua obra* (Lisbon: Fundação Calouste Gulbenkian, 1986).

44 Antonio Menéres, "Evocação, Lúcio Costa, A liçao do arquitecto," *Jornal de Letras*, July 1, 1998. Menéres recalled experiencing, on his first visit to the Palácio Capanema, "that thing *mestre* Ramos always talked about in his lessons: the integration of arts."

45 Among them Viana de Lima, José Carlos Loureiro, Jorge Pais da Silva, Arnaldo Araujo, Augusto Amaral, and Manuel Ferreira.

46 *Arquitectura popular em Portugal* (Lisbon: Sindacato Nacional dos Arquitectos Portugueses, 1961).

47 Costa, *Registro de uma vivência*, 457.

48 *Arquitectura popular em Portugal* (Lisbon: Associação dos Arquitectos Portugueses, 1988).

49 Some letters sent by the architect between the mid-1960s and the early 1970s (in the CLC) testify about other Portuguese commissions: one, addressed to his daughter Maria Elisa, concerns a large residential complex in Odivelas, and a non-identified project for the Quinta do Tejo; in another one, Costa refuses to contribute to a public competition for the urban development of Cascais, indicating the name of the Portuguese architect Viana de Lima, among those who accompanied him in 1961 on the visit to Trás-os-Montes.

50 This was the third competition for Sagres and the first one open to foreign contributions. See *Carlos Ramos, Exposição Retrospectiva*. According to José Pessôa, Costa did not send his proposal to the attention of the jury.

51 Lucio Costa, manuscript draft of the written project proposal for the Sagres monument, ANS/IPHAN. Notice that this formulation anticipates the rhetoric used some years later by Costa in the introduction to Brasília's Plano Piloto.

52 "Mar novo. Sintese Alada do Genio Portugues das Descobertas," *Padrão* 55 (December 1956): 31–33.

53 See Felipe de Souza Noto, "Plástica e Tradição: uma aproximação entre Lucio Costa e Fernando Távora," http://www.docomomo.org.br/seminario%206%20pdfs/Felipe%20Noto.pdf.

54 Costa, *Registro de uma vivência*, 226.

2

"CORBUSIANS" IN URUGUAY

A Contradictory Report

Jorge Nudelman Blejwas

The presumed overriding influence of Le Corbusier in Latin American architecture has produced a distorted view of his actual influence in the period prior to the Second World War. Of late, this viewpoint has been reinforced by the pageantry around the 1929 visit, which continues to generate texts on a trip that had no significant outcome in terms of the architecture of the countries he visited.[1] Certainly it had an effect on the visitor himself, but this has been analyzed at length.[2] Neither did the eloquent sermons of Le Corbusier produce masses of instant adherents in Uruguay, and those exceptional cases in which the Swiss master's theoretical directives were followed and his formal codes explored are clearly evident. In practice, these are limited to a list of direct collaborators and a handful of followers. Le Corbusier's key Uruguayan disciples were Carlos Gómez Gavazzo (in the atelier for five months in 1933–1934), Justino Serralta (1948–1951), and Carlos Clémot (1949–1950). It seems easy to connect the first visit in 1933 with Le Corbusier's 1929 trip, but a more detailed analysis allows us to appreciate that this was not in fact such an obvious choice for Gómez Gavazzo. On the other hand, the visits of Serralta and Clémot were; with the war just ended, Le Corbusier appeared as a sound European alternative to the ascendant trends coming from the United States.

Mauricio Cravotto and the Rise of the Architect

In 1929, Uruguayan architectural culture was characterized by a Beaux-Arts methodology that facilitated eclecticism, with forms moving between the classical, the nostalgia for a scarce "colonial" heritage, and modern tendencies, including Art Deco. Architectural education had established the foundations of its French tradition under the direction of Professor Joseph P. Carré from the École des Beaux-Arts in Paris, appointed to the Universidad de la República in 1907, where courses on architecture were taught within the Department of Mathematics. "Monsieur Carré" earned enormous prestige, based on his quality as a teacher and his tolerance of—even drive toward—innovation. Almost mythified following his death in 1941, his prestige persisted despite the changes that came to architectural education after 1952. Up until then the educational program was centered on the "Cátedra de Proyectos," the design course organized in two cycles, followed by a postgraduate course in Grand Composition, leading to the Grand Prize and its travel scholarship for overseas study. In addition, the courses on urbanism were independent and, if we include the Construction Project and Decorative Composition courses in all their various levels, the result was a fragmentary system that assumed an eventual synthesis. Decorative Composition consisted of the solution of architectural problems with an emphasis on expressing the essence of a theme. Exercises with a strong didactic intent were proposed: plazas and monuments (in the most traditional sense), but also theater and film sets, or subjects related to advertising. Despite its academic roots, it was a fertile space for experimentation.

The assertion of a professional identity for architects was greatly supported by the Sociedad de Arquitectos del Uruguay (SAU) from its foundation in 1914, particularly through its official journal, *Arquitectura*; this coincided with the establishment of a dedicated Department of Architecture in 1915. The training of architects and engineers was separated that year, and a tough battle ensued over the definition of their respective professional expertise, in which the architects insisted on their ability to resolve symbolic and representative questions through art. This claim coincided with the dissemination of modern architecture, first received through magazines, and later through more direct sources following the travels of scholarship recipients such as Mauricio Cravotto, who was the first to travel overseas and is recognized as introducing modern tendencies, which at that time were generally perceived as a homogeneous whole. His "grand tour" would take three years (1918–1921), traveling to the United States, England, Spain, and Italy, before settling in Paris, where he attended the classes of León Jaussely, Marcel Poëte, and others, to whom he would always acknowledge his debt. In 1936, he created the Institute of Urbanism within the Architecture Department, along "scientific"—and eventually anti-Corbusian—lines. In 1938, in his report on a trip he had recently made to Italy and Germany (and which has since been systematically erased from Uruguayan historiography), he declared:

> The marked aversion that for some years I have felt toward ultra-civilized modern cities has led me to select my centers of study, markedly preferring the lesser urban centers which, perhaps by being small and perhaps by being pressed by geographical demands, are outstanding organisms that are as harmonious and complete as a beautiful product of Nature.[3]

Although anchored in this romantic line of thought, his first fascination with modern architecture has encouraged a hasty reading of Cravotto as an avant-gardist, which is contradicted by his few known writings, and by his love of Italian Gothic. Even so, Cravotto had cultivated the friendship of Alberto Sartoris, who regularly mentioned Cravotto in his books. In parallel, he developed a certain German addiction (which included the Austrian and Swiss) as an admirer of their artists and philosophers, whom he quoted continuously. Nevertheless the 1938 trip hints at a curiosity that borders on the political, and that coincides with the anti-metropolitan sensibility of his report. Despite these ambiguities, Cravotto would become the catalyst for a group of architects who felt at once committed to progress and concerned with "identifying an identity" that connected them to a worthy past. Art and architecture were to be keystones in generating these links between past and future, and architects would fight for their place in constructing the image of the nation. This professional assertiveness was particularly notable in the 1930s and 1940s: an extraordinary leadership in the social and political realms enmeshed architects with the government and the state. The proliferation of media outlets (four—and sometimes more—professional magazines published concurrently, exhibitions, radio broadcasts, etc.) was consistent with this moment of euphoria. From this period comes the formation of a canon of Uruguayan architecture—successively enriched—that is inhabited by projects with a strong inclination toward modernity even though heterogen-eous in style. Despite this, the Beaux-Arts methodology lived on: *parti, esquisses,* grand compositions, the "decorative," the theory of Julien Guadet, the historic rationalism of Auguste Choisy.

Against this French background, Spanish, Italian, and German contributions—all gathered together in *Arquitectura*—were counterposed and integrated. In fact, the Spanish press was the first to herald modern architecture from Germany, though soon after the original literature was drawn upon. The Dutch influence was also powerful, and a touch of the Italian was visible in the built work of some successful architects. There was a notable convergence with the German and Italian sympathies of those in power following President Gabriel Terra's coup d'état in 1933, and even more with his successor, the "General *Arquitecto* Don Alfredo Baldomir," as *Arquitectura* celebrated him in 1938,[4] while highlighting the fact that the Mayor of Montevideo and the heads of three ministries were likewise architects. Unquestionably the social prestige of the profession reached its zenith at this moment. This is evident in both the quality and the quantity of public works projects, in a nation that since the early twentieth century had been enacting strong policies in support of state control.

In 1932, the magazine of the Centro de Estudiantes de Arquitectura (CEDA) joined the landscape of Uruguayan architectural culture. The first six issues, up to 1934, expressed populist political leanings, and a certain leftward slant in the tradition of Latin American universities (identified with the university reform movement of 1918 in Córdoba, Argentina), as well as a clear opposition to Terra's 1933 coup. We see articles by and about the Mexican muralist David Alfaro Siqueiros, about socialism and declarations against fascism, as well as students' harsh critiques of their professors, among other academic demands. The counterbalance of Cravotto is evident: when invited to participate, he systematically nominated German authors, providing the translations himself. In 1936, *CEDA* retreated from its more radical positions: with changes in management it became more conservative, and the support for new architecture waned. Publication became irregular, and although protests continued, they were now politically cautious and superficially focused on the profession. We have to wait until September 1940 to find an editorial alluding to the war. This stage of *CEDA* was brought to a close by another push from Cravotto: in May 1941, he translated an article by Peter Behrens and wrote an introduction for one by Werner Hegemann. Nevertheless, the presence of certain themes gradually increased: the curriculum, university reform, and questioning certain professors, which were debated with vehement frankness. In 1942, in tune with the new direction of the war, even the titles of the articles perfectly illustrated the spirit: "New Blood," "Toward a New Department," and "To the Fight!"

But we mustn't get ahead of ourselves.

Germans, also in Montevideo

One of the observations stemming from Le Corbusier's visit to the Río de la Plata—which not all critics have picked up on—was the quiet complaint he let slip concerning the influence of the Germans in Montevideo.[5] Le Corbusier wrote to his mother:

> Spent 3 days in Montevideo. Delightful, joyful. . . . And I was received by the Students and their professors all full of life and of joy. And I was received as "the man of the moment!" . . . A city fervently devoted to the Latin spirit. But the Germans are making a huge propaganda effort here.[6]

A messiah, but against the backdrop of German domination. Le Corbusier's Uruguayan admirers did not notice the discourtesy. For them, the modern was, from a distance, a worldwide phenomenon, albeit dominated by German-speaking theoreticians. Cravotto's influence over the Architecture Department's library may be significant here, as was his enthusiasm for the pedagogical value of the tours, which ended up giving him a leading role in the selection for the Grand Prize—he usually drew up the rules for the contest—and in the initiative of organizing group tours for students beginning in 1947.

The accounts of the 1929 visit are eloquent concerning the weight of all things German in the Uruguayan academy (note how out of date the French authors were, in contrast to the contemporaneity of the German names):

> the conversation was enlivened upon touching the subject of drafting, and upon sounding the names of Hegemann, Camillo Sitte, Viollet-le-Duc, [Ernst] May, Walter Gropius, Marcel Mayer, Haussmann, Eiffel, Monier the inventor of reinforced concrete, and Frugès of Bordeaux.[7]

Because of this, Le Corbusier took care to establish some distance for himself, underscoring his differences with the Germans and his conflicts with Peter Behrens, perhaps perceiving the students' empathy with leftist posturing:

> After reviewing the new architects and urbanists, Le Corbusier had misgivings about the work of [Hermann] Jansen and that of Peter Behrens, with whom he worked when starting out as a draftsman. . . . He spoke of his enmity with Peter Behrens, born out of a strike by that architect's staff. Le Corbusier, disinterested defender of the workers to whom the afore-mentioned architect paid starvation wages, organized and led the protests that became an energetic and tenacious strike.[8]

Further on he regretted his "Germanophobia," which he apparently overcame.[9] Despite these low blows, Le Corbusier's pretense of breaching the German divide, however, was not fruitful: the Uruguayan-German links were already too close. As we will see, it was only by accident that Gómez Gavazzo chose to stay in Paris instead of maintaining his original German itinerary.

Le Corbusier was taken to see various projects in Montevideo—clearly already modern—and visited several architects' studios. In the studio of Juan Rius and Rodolfo Amargós, he saw the first version of the Facultad de Odontología, designed under the influence of W. M. Dudok. Despite the myth that the second and definitive version was influenced by the French visitor, the final project is more German than Dutch, closer to Neue Sachlichkeit abstraction than to Corbusian Purism. Le Corbusier recalled the 1925 visit to Paris made by Rodolfo Amargós during his Grand Prize scholarship tour. However, Amargós had not been on a pilgrimage to Paris; he had only been passing through. In fact, he had written from Vienna: "Strongly attracted . . . by this newest trend—which I judge to be most interesting for the spirit of our young race—with advanced principles and devoid of all national artistic tradition, I applied for a place in Arch. Behrens workshop."[10]

Similarly, scholarship recipient Rosendo Quinteiro slipped into a report from Paris that he was learning German in order to "travel through the Nordic countries, creators of the modern architectural movement."[11] Quinteiro traveled to the Netherlands, and then, after a trip through Stuttgart, Munich, Cologne, Düsseldorf, Essen, Leipzig, and Dessau, in November 1927 he arrived in Berlin where he arranged to study architecture with Hans Poelzig and urbanism with Hermann Jansen.[12] Having passed through Dessau, he must have visited the Bauhaus, but he chose Poelzig. Subsequently, Juan Antonio Scasso's 1932 tour, using a "faculty enhancement scholarship," confirms the persistence of the German line. Scasso was municipal architect in Montevideo and second to Cravotto at the Institute of Urbanism. His tour objectives were to study: "A) Green space policy; B) Housing policy; C) Urbanism in Holland; and D) The work of an architect: Dudok in Hilbersum."[13] His itinerary was also significant: he traveled through Germany from mid-June to mid-August 1932, interviewing municipal architects and visiting the Siedlungen. After traveling through Holland for only nine days (sadly, Dudok was on vacation), he arrived in Paris on August 23, but this would only be a courtesy call: "In Paris I immediately went to visit *Arquitecto* Le Corbusier." Le Corbusier then put Scasso in contact with Josep Lluís Sert in Barcelona, while in Madrid Secundino Zuazo Ugalde and Fernando García Mercadal would be his hosts.

It is no surprise that in 1929 Le Corbusier revealed his private disappointment in the letter to his mother. Despite the warm send-off that he was afforded as a hero of modernity, he knew that he had arrived in Montevideo too late.

Yet all was not lost.

Carlos Gómez Gavazzo and the Shift toward Paris

Carlos Gómez Gavazzo has consistently been identified with Le Corbusier in Uruguayan historiography. He earned his architecture degree in 1932, and that same year he won the Grand Prize. The "signs of modernity" in Gómez were precocious and well recognized. In 1927—while still a student—he designed a house for Valerio Souto, which would become one of the most radical expressions of modern architecture in Uruguay, without the decorative contaminations that were still evident in contemporaneous works. Despite being the work of a student,

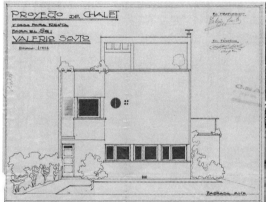

2.01 (above) Carlos Gómez Gavazzo, Valerio Souto House, Montevideo, 1927

2.02 (below) Carlos Gómez Gavazzo, Cavia Street Houses, Montevideo, 1927 it was included in Le Corbusier's 1929 itinerary. The ability to appropriate the modern manner for himself and to still experiment with it (also seen in the lesser-known houses on Cavia Street, from 1930) placed Gómez on the same level as his masters.

The main destination on his scholarship tour—despite the recent visit of the "messiah"—was again Germany, "where the undersigned [Gómez] wishes to complete courses in urbanism, after visiting Spain, Algeria, Italy, Austria, and France."[14] Having traveled through Spain and Italy, in Paris he requested an interview with Le Corbusier, and was invited to enter his studio. At the end of the summer, he started working at 35 rue de Sèvres, remaining there until February 1934. During these five months, Gómez participated in the second version of the Algiers project and the design of the Rentenanstalt in Zurich.[15]

His letters to Montevideo suggested a great deal of leadership: "I am pleased to let you know that I have started to work under the direction of *Arquitecto* Le Corbusier, taking charge of the study . . . of the master plan for the city of Algiers."[16] (Let us note in parallel: in August 1933 Le Corbusier was on the *Patris*

2.03 Carlos Gómez Gavazzo, drawing of Piazza San Marco, Venice, 1933

EL CONJUNTO DE SAN MARCOS DIBUJO DEL ARQT. GOMEZ GAVAZZO

C.E.D.A. CENTRO ESTUDIANTES DE ARQUITECTURA 6

II for the CIAM Functional City congress, heading for Athens; traveling with him as part of the Spanish delegation was Antonio Bonet, still a student. In 1945, Gómez and Bonet would compete over the plan for the Punta Ballena subdivision.) In September, Gómez translated the conclusions of CIAM 4—obviously provided to him by Le Corbusier—sending it to Montevideo, although it was never published. Having concluded his French sojourn, a trip to Germany was still mandatory before returning home. Nevertheless, the excursion lasted a mere ten days.

Gómez's first public appearance upon his return to Uruguay was in the *CEDA* magazine, with five travel sketches and a text describing his first visit to 35 rue de Sèvres and the atmosphere there. Let us put aside his words for a moment and examine some of his drawings from this issue of 1934. The magazine cover contains an aerial view entitled "The San Marco ensemble." The drawing has a peculiar feature that almost passes unnoticed: Gómez has subverted the buildings' facades, drastically simplifying them, converting them into modern blocks with *fenêtres en longueur*. In Venice, even before arriving in Paris, he appears to be rehearsing the modern through (of course!) a change of masks. Inside we find two additional drawings, more traditional in execution: a sketch of the duomo and the campanile of Pisa. Then, accompanying his narrative about Corbusier's atelier, is an analytical drawing of the Pantheon in Rome, dated June 1933, with handwritten observations. Finally, there is a Parisian perspective on a double-page spread, with the following reflections:

> *Space is the first requirement of the modern city*. Population centers become congested in population and intensity. Hygiene and modern communications media create spaces of gigantic proportions. Speed measures distances that are shortened in inverse proportion. The new space that is created is a volume that is measured in 3 directions; the verticality of old structures is insufficient to form a volume with the large planes that are created. *It is necessary to give proportion to the frame*, with the necessary vertical element that our era and our means demand. *A change of scale is stressed*. An aesthetic that is *new in its magnitudes* begins. . . .
>
> The Eiffel Tower measures the volume of the Champ de Mars—it is the 3rd dimension that gives the space its scale. Completes the vision, and announces the conception of new compositions. *This element of height is wanting in order to complete the volume (element of architectural aesthetics) in the Champs-Élysées.*[37]

How far from the Plan Voisin these notes seem, at the doors of 35 rue de Sèvres. The reflection (it cannot be described as a proposal) on Paris and space in the modern city sets out from its own icons. The Eiffel Towers that Gómez extrudes in parallel are not, for those who wish an "architectural" explanation, buildings with any function. They are theoretical volumes constructing the space. For Gómez, the problem is not to build the modern city, but to find the most suitable form for the city that is already modern. His preoccupation is not functional, but aesthetic, textually. The trial sketch from Venice reinforces this. The spatial and architectural structure of San Marco is untouchable—it was one of Cravotto's obsessions—yet it is possible to experiment with it. How much is here of the modern, in the sense of a "tabula rasa"? In view of the drawings, very little. It is likely that they are from the summer of 1933, before starting his *stage* (or training period) with Le Corbusier.

The account of his arrival at Le Corbusier's studio is written in the form of a dialogue, perhaps not very informative, but appealing. Beyond its value as testimony, the scene is recalled as a workshop rich in exchanges between apprentices.

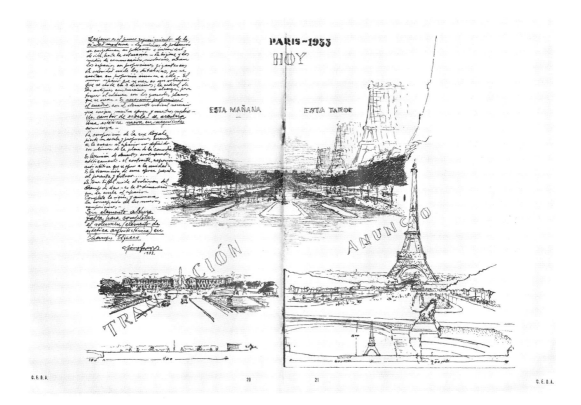

2.04 Carlos Gómez Gavazzo, drawing of Paris streetscape, 1933, published in *CEDA* 6

It is easy to understand the drive to excel that develops in this work community; very often giving rise to interesting conversations on architectural concepts and realizations; true polemics, sometimes aggravated by the difficulties that, in such moments, this small Babel presented. But if one language fails to bring understanding between a Japanese and a Russian, and between a Russian and a Czechoslovakian, a Swiss, and a Uruguayan, the sketch is always more valuable than the word, because besides expressing, it specifies and defines. This working in common is no secret to students, as much as it represents stimulus and benefits that contribute to their interests, besides encouraging a camaraderie, which is strengthened for a lifetime, with practical results too.[18]

In 1943, when presenting his thesis in a competition for a faculty position as professor of design in the Architecture Department, these recollections became a manifesto. Gómez proposed a possible alternative to the fragmented teaching of design (at that time still closely modeled on the École des Beaux-Arts) through the unification of all the design studios into one vertical workshop, a single space where all architecture—at all scales—would be discussed, analyzed, and designed. In fact, a kind of "builders' lodge" (or *Bauhütte*) presided over by a master while the apprentices—hierarchically organized—instruct one another. Gómez unequivocally defended the academic tradition he evidently perceived as under threat:

> Just like that the *esquisses* went away and the classical examples were lost; then faith in the professorship was lost and even the corrections given by our great Master [Carré] were disputed, putting in doubt his eloquent wisdom—I was a witness to this.[19]

Reclaiming classicism and *esquisses* in 1943 appears quite striking. However, it is less so if we view it against the background of the order and mathematical precision of a persistent Corbusian Neo-Platonism and Neo-Pythagoreanism in Gómez. Yet nonetheless, other ingredients are superimposed in Gómez's proposal, particularly a medievalism and scientism deriving from Cravotto—an internally contradictory mixture that would end up dividing the faculty, even to this day.

The architects' professional magazine would also give Gómez some space—perhaps not a very generous amount, since he was becoming a figure of dissent—for articles on architecture and urbanism, and eventually some projects. In 1937, *Arquitectura* published his competition proposal for the layout of Avenida Agraciada, in Montevideo,[20] to resolve the intersection between the city's main avenue and the view toward the neoclassical Legislative Palace, inaugurated ten years earlier. The relationship with the Plan Obus for Algiers—or at least with the

2.05 Carlos Gómez Gavazzo, competition project for Avenida Agraciada, 1937

reduced version of Plan B, on which Gómez worked—is explicit. But how much of Gómez is there in Plan B for Algiers? How much of that incipient monumentality, or of the building's symmetry with the edge of the viaduct crossing the thin tower block, could belong to an architect obviously preoccupied with a Germanic architectural "expressionism" learned at the French-styled school in Montevideo? If Le Corbusier had been inspired to loosen and twist the Cartesian volumes in the first version, achieving an organic yet still abstract form, in Plan B—less widely disseminated—the axes of symmetry and academic sympathies reappear. The deviation toward the monumental would be somewhat corrected in the third version (Project C), which is limited to the grand edifice of the Cité d'affaires, and where one notes the absence of chiaroscuro from a hand trained in the Beaux-Arts (such as Gómez's).

Between 1935 and 1956 Gómez participated in numerous competitions, generally with little luck. In 1935, he submitted an unorthodox project—evidently indebted to his Paris sojourn—to the competition for the Uruguayan Cámara de Comercio (Chamber of Commerce); the first-placed project being a symmetrical and sober unornamented building.[21] In Gómez's unsuccessful proposal, all of the symbolic gestures are in the attached sculptural monumental forms, which are clearly molded and separated from the architecture. Details such as a meeting room in the form of a truncated cone culminating in a cupola, a ground floor that opens up generously into public space, even a structure that is somewhat hard to determine, make this a "difficult" project. The building, far from assuming the conventional form for this typology, creates a stage for itself, and exhibits an exoticism (it is, to a certain extent, Brazilian) that was totally extravagant for the times. It was ignored by the jury.

In 1938, the competition for the Uruguayan Palacio de Justicia was declared void, with no first prize awarded. In this case, Gómez raised the stakes with a project that brushed the edges of its own alleged rationalizations (bringing sunlight to the narrow back-alley, and creating an image that clearly spoke of justice): comprising a prism that ensured the sun's reach and a traditional monument (in the most banal sense of the word), the project nonetheless avoided the then-

2.06　Carlos Gómez Gavazzo, competition project for the Uruguayan Cámara de Comercio, 1935

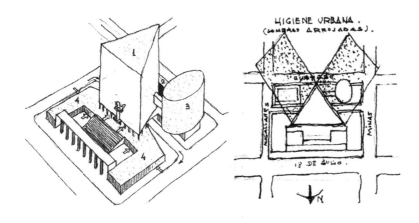

2.07 (left) Carlos Gómez Gavazzo, competition project for the Uruguayan Palacio de Justicia, perspective and shadow studies, 1938

2.08 (below) Carlos Gómez Gavazzo, competition project for the Palacio Legislativo in Quito, 1945, regulating lines diagram, as drawn for publication in 1973

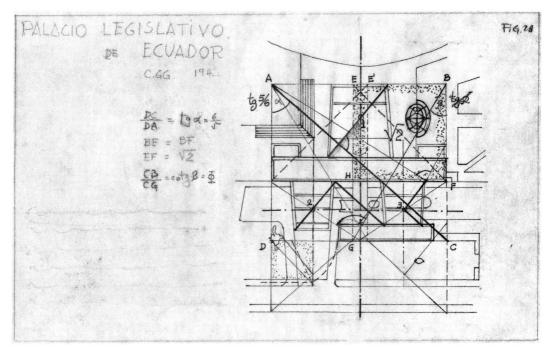

popular purified "modern classicism" recurring in all of the projects given awards. This was a juxtaposition impossible for Gómez to synthesize, even though it suggests a common root with Le Corbusier, who would subsequently pursue a similar effect with greater success—particularly with the Open Hand in Chandigarh, as well as Marseilles and Ronchamp.

In 1945, Carlos Gómez Gavazzo was honored with second prize in the first phase of the competition for the Palacio Legislativo in Quito, Ecuador.[22] A year later *Arquitectura* presented the project in its issue dedicated to the VI Pan-American Congress of Architects in Lima: it had already become part of the canon of Uruguayan architecture.[23] In 1948, the magazine dedicated its main section to the second phase of the competition, which Gómez had just lost. Lavishly illustrated and documented, it culminated with an open letter in which Gómez courteously protested against the regionalist criteria of the jury, calling upon it

2.09 (right) Carlos Gómez
Gavazzo, competition
project for the Palacio
Legislativo in Quito, 1945—
Gómez Gavazzo's
collaborators constructing
the model

2.10 (below) Carlos Gómez
Gavazzo, competition
project for the Centro de
Asistencia del Sindicato
Médico del Uruguay, 1949

"not to receive as FUTURIST conceptions that endeavor to highlight with a contemporary spirit the eloquent and imperishable structures of San Francisco."[24] Quito was, for the architect, Serlio's stairs for the Church of San Francisco, which are included in his evocative collage (the anachronistic meaning of the term must be excused); moreover there was a rigorous discipline in his effort to introduce the project into Quito's geography. In this way, Gómez offered bridges toward the past; as his letter reaffirmed: "Past and present should be united in a harmonious bond, interwoven by the subtle threads of a composition that is eternal, because it is in this delicate play where nature's most vibrant lights are deciphered."

The 1949 competition for the Centro de Asistencia del Sindicato Médico del Uruguay illustrates the shift in direction of Uruguayan architecture over the next twenty years.[25] The first prize would go to two disciples of Gómez: Francisco Villegas Berro and Alfredo Altamirano. Their teacher won the second prize, with a proposal that was highly consistent with his earlier concerns over the legibility of architecture, now coinciding with the emphasis that Le Corbusier himself was placing on communication and monumentality. But the giant torch, symbolizing the medical profession, placed on the facade's axis of symmetry, was very far from the rough monumentality à la Léger that Le Corbusier adopted beginning in the 1940s. Gómez still trusted in the literal value of the image; for this reason, the separation between architecture and symbol was radical.

Gómez's architecture was iconographically Corbusian. He resorted to the trapezoid from the plan of the Palace of the Soviets time and time again, as in the competition for the Biblioteca Nacional de Uruguay in 1937; we can also recognize the principles of "pure volumes" in the competition for the Palacio de Justicia. However, when the building must "speak," it does so literally, leaving the architecture aside, through the method of adding allusive ornamentation. In this respect, it is useful to draw a comparison with the metaphorical monumentality that would soon appear, in the architecture of Brasília, for example. Brazilian architecture—perhaps much more than any other—was infused with Le Corbusier, but Gómez's strategies are prior, more academic, even "Loosian"; the necessary symbol is always added to an architecture that, despite lacking the indifference of the Viennese master, knows itself to be equally mute.

After 1945, trends from the United States and Brazil would be the new models, leaving aside the collection of modern classicisms, Art Deco, or the *littorio* style dominant in the interwar period. This was aided by the lull in European architectural production, but also by trips to South America and the United States by recipients of the Grand Prize and other scholarships, who had been unable to travel to Europe. In this context, the recovery of Wright is notable, as well as the fascination with Neutra, and the discovery of the US "Miesian" line.

Mauricio Cravotto continued to persist with demands anchored in his medievalism: introducing Richard Neutra (who visited Uruguay in 1945), he forcibly inserted his anti-Corbusian and Manichean humanism with a final paragraph taken from historical geographer René Clozier: "the architects of the Middle Ages were not mass-producers of machines for living in, but rather good builders and good decorators, at the same time; that is to say, architects."[26] It was a late, doomed offensive, despite the pyrrhic victory of the competition for the Master Plan of Mendoza (1940),[27] where Cravotto and Juan Antonio Scasso, together with their Argentine collaborators Fermín Bereterbide and Alberto Belgrano Blanco, won the first prize, leaving the architects of Grupo Austral (supposedly led by Le Corbusier)[28] in third place. This was a triumph, certainly, not only for the conservative urbanism of its authors, but also for Carlos Maria della Paolera, who had studied with Marcel Poëte in the 1920s and had become an obstructionist of vanguardist positions, very much allied to a realism closely shared with Cravotto and their mutual friend, Werner Hegemann.

Around 1952, other battles would follow. As a result of Gómez's persistent calls for reform, and of the enthusiasm of students already aligned to the Marxist left following Germany's defeat, the Architecture Department reformed its curriculum. A tacit alliance was formed between the student leaders (mostly Trotskyist) and Gómez: while he drafted the main academic content of the study plan, they contributed the social concerns outlined in their Mission Statement. On the pedagogical level, the unification of the design professorships—including urbanism—is notable, along the lines Gómez had advocated in his 1943 thesis. (It should be remembered that he had been, simultaneously, professor of Design, Decorative Composition, and Urbanism.) The consolidation of all the design studios under one sole chair was, in structural terms, an act of nostalgia, perhaps an unconscious return to the Gothic *Bauhütte* for the training of master masons. It is difficult to establish a direct link—there are few mentions of Gropius and Giedion in the Mission Statement—but these are precisely the same foundations as the first Bauhaus, reaffirmed by the Paris memories that Gómez had recounted in the pages of *CEDA* almost twenty years earlier. From a fragmented conception of education that forced the student to make connections, and that encouraged contradictions among the different branches of the design studio, the "Plan of '52" wanted to affirm the unity between technique, expression, and concept.

While these changes happened, other Uruguayans traveled to Paris.

Serralta and Clémot, and the Path toward Dieste

Carlos Gómez Gavazzo was one of the four architects nominated by Le Corbusier to direct the construction of the Curutchet house in La Plata, Argentina.[29] Let us note that the recommendation came in 1948, fourteen years after his return home, but coinciding with the presence of Justino Serralta in Paris. Serralta was another Uruguayan disciple who, along with Carlos Clémot, worked with Le Corbusier between 1948 and 1951. Moreover, it was they who would convey to Eladio Dieste the basic principles of a certain Corbusian doctrine, drawn from the Modulor, along with other prescriptions that he would adapt for his best-known works. In collaboration, the architects (Serralta and Clémot) and engineers (Dieste, along with Eugenio Montañez) worked together on—and took equal credit for—a number of projects between the mid-1950s and the end of the 1960s.

In 1947, Serralta—although by then already a graduate—joined the first group tour of students from the urbanism program, which traveled to Italy, Switzerland, France, and Spain under the sponsorship of Mauricio Cravotto. He left the group in Paris, to eventually join Le Corbusier's studio, where he remained for three and a half years.[30] Initially, he worked on the infrastructure for Marseilles, but later he was able to report that he had been "entrusted with the study for the Terrace of the Unité d'habitation, where I had the opportunity to be in direct contact with L.C. in the creative order."[31] He also fabricated the plaster model of Ronchamp along with André Maisonnier and, according to his own account, just before returning to Uruguay Le Corbusier offered him the opportunity to work at Chandigarh. Concerning Carlos Clémot, we know only that he was in the atelier between 1949 and late 1950; the two Uruguayans would have become friends in the master's studio and, presumably, plotted the strategy for their return while they were there.

Parallel to his work on the Unité, Serralta developed the Modulor with André Maisonnier, producing a series of drawings that Le Corbusier would utilize and discuss extensively in *Modulor 2*, published in 1955.[32] This marked the first appearance of a drawing acknowledged by Le Corbusier as Serralta's invention—which Serralta (apparently) first baptized "square 1, 2, 3, 4, 5" and then later "tetrator"—a geometric principle based on the Pythagorean "tetraktys" drawn from a square. This was a key element of Serralta's later theoretical development: the *Unitor*.[33] But why a *Modulor 2* after the effort of the first? If a Modulor had to be sent to Milan in 1951 ("Installed in a beautiful spot in the Exhibition on 'Divine Proportion' at the 1951 Milan Triennale, in the company of the manuscripts or first editions of Vitruvius, Villar[d] de Honnencourt, Piero della Francesca, Dürer, Leonardo da Vinci, Alberti, etc., etc."),[34] it is possible that in 1950 Le Corbusier had entrusted Serralta and Maisonnier with the task of making the sketch. This is the drawing that illustrates the cover of *Modulor 2*, whose authorship Le Corbusier attributed to the pair, describing it as a "historic" achievement. Figures 15 to 18 of *Modulor 2* are also drawings by Serralta, easily recognizable by his touch and his handwriting, besides the comments of Le Corbusier, which reveal an exceptional closeness and affection: "Here is a drawing (fig. 15) provided by Serralta and Maisonnier: they have taken the square of 'the 1.83 m. tall Modulor man' (but since Serralta has a tender heart, his man is a woman 1.83 m. tall. Brrr!)"[35]

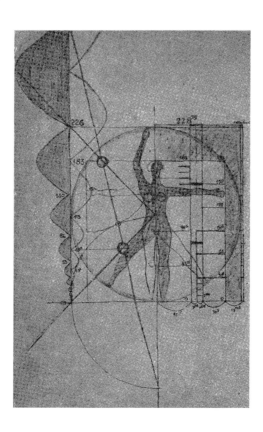

2.11 Justino Serralta, "Femme modulor," as reproduced in *Modulor 2*

This success and recognition would mark Serralta; the focus of his theoretical inquiries became obtaining an abstract rule for the order of the universe, with Pierre Teilhard de Chardin (Le Corbusier's suggestion, later conveyed to Dieste) as his key point of reference. Concerning the dilemma of accepting Le Corbusier's offer to go to India or returning to the Río de la Plata, the security of home proved stronger. Serralta's role in proximity to the master must have seemed sufficient credentials to distinguish himself on the Uruguayan scene; this was confirmed by his rapid inclusion into the Taller Altamirano—a prestigious modernist atelier within the Architecture Department—as well as the Institute of the Theory of Architecture and Urbanism, along with Carlos Gómez Gavazzo (note its change of name: formerly the Institute of Urbanism, now architecture and urbanism were to be considered within one seamless framework).

The scene had changed radically in Serralta's absence. In 1947, the Department had moved to its newly finished building, and the new curriculum was put into practice in 1952. The expulsion of Cravotto from the renewed Architecture Department was a sign of the times. The traces of the machinations are evident in his letter of resignation:

> With good spiritual disposition and good health, I was awaiting the beginning of teaching duties on February 2nd, when by a communication from the CEDA . . . I learned of the resolution declaring a general pre-strike by the students, citing among other reasons, "the violation of the curriculum" on the part of the Directive Council of the Department by reelecting me as professor of design . . . and also telling me this: "activating the student body's struggle for the essential principles of the curriculum and the cleansing of the faculty."[36]

He would not be the only one expelled. With these events, the most "academic" period of the Department of Architecture came to a close.

Construction and Utopia

Among the papers in the Serralta and Clémot archives are materials from Le Corbusier's studio, including two copies of the designs for the vaulted houses of the Permanent City at La Sainte-Baume:[37] what draws our attention is their iconic power. The design solution—a series of parallel barrel vaults, better known as the "Roq et Rob" project—would be imitated by Serralta and Clémot time and time again. None were ever built; it would not be fruitless to analyze the influence of engineer Eladio Dieste in this repeated failure. A small pencil drawing of a typically Diestian Gaussian vault, made in one corner of their copy of the floor plan of La Sainte-Baume, could be interpreted as the engineer's contribution to the discussion over a more rational and economical way to build vaults.

In 1947, Dieste had successfully resisted Antonio Bonet's efforts to build the vaults of the Casa Berlingieri in Punta Ballena out of reinforced concrete. The original idea of building Catalan vaults had failed early on, although the invention of "reinforced ceramic" would emerge from this idea. Le Corbusier's requests for advice from Domingo Escorsa Badia on how to build the roofs of the Maisons Jaoul using the same masonry vault technique—as documented in the *Œuvre complète*—makes for a remarkable coincidence.[38] In the end, the construction of the Maisons Jaoul by Algerian workers (although in contemporary writings they were persistently denominated "Catalan"),[39] illustrated the vernacular leanings of the architects, as opposed to the technological and economic rationality of the engineer. In the end, Dieste moved toward reclaiming an ancestral wisdom in his texts, with frequent references to the craft of building, and to manual skill merging with the dictum of theory (with himself as spokesperson). But in any case, this reclaiming of tradition directed him away from any attempts at folkloric imitation.[40] Instead, his efforts focused on incorporating handcrafted materials, skilled foremen, and sophisticated calculation and procedures into the same process. For Dieste, the form was a consequence of this quasi-natural process, out of which a new tradition would arise. Nonetheless, Serralta and Clémot insisted that the barrel vaults of the housing for La Sainte-Baume would appear in almost all of their projects from their years working with Dieste. In the Colegio La Mennais project, started in 1958, these were—by the final version of 1965— reduced to a pair of conical vaults facing each other in the small chapel on the upper floor.

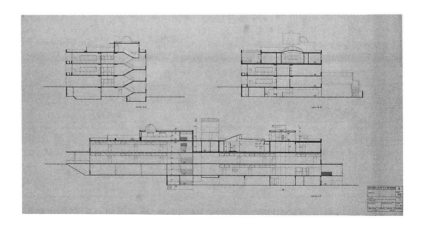

2.12 Justino Serralta, Carlos Clémot, Eladio Dieste, and Eugenio Montañez, sections of the final project for the Colegio La Mennais, 1963

2.13 (left) Justino Serralta with two priests in the chapel under construction at the Colegio La Mennais, ca. 1964

2.14 (below) Carlos Clémot and Justino Serralta on a roof under construction at the Colegio La Mennais, ca. 1964

The vaults also appeared in unbuilt projects, such as the house for engineer Eugenio Montañez, or in housing blocks that were closer to the Marseilles model, or even warehouses—in which, finally, Dieste's innovations prevailed.

Serralta and Clémot began to use folded plates of reinforced ceramic tiles, and—advised by Dieste—experimented in the Casa Acosta y Lara (1961) with overhangs made of reinforced ceramics. The most complete expression of this experimentation was, without a doubt, Dieste's Church of San Pedro in Durazno, whose Corbusian affiliations have already been established.[41] Dieste's key works had already acquired a solid reputation locally and regionally. This produced an unexpected result: engineer Eladio Dieste began to be considered the most original "Uruguayan architect"—even appropriated by Uruguayan architects themselves. A paradox, if one takes into account the efforts in those first years of the twentieth century to establish separate and specific professional identities for architects and engineers.

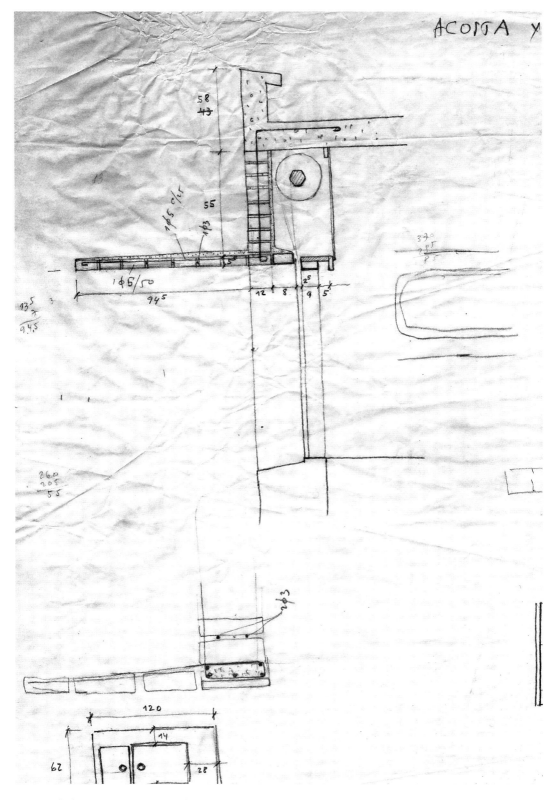

2.15 Detail of window *brise-soleil*, in reinforced brick, designed by Eladio Dieste, drawing ca. 1963. Justino Serralta and Carlos Clémot, Acosta y Lara House, ca. 1961

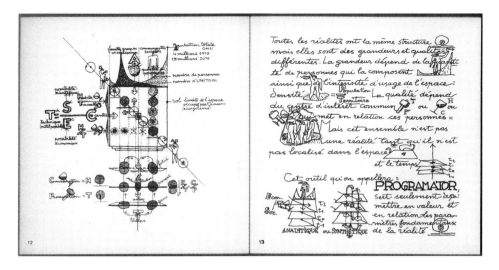

2.16 Justino Serralta, "Le programator," *L'Unitor*, 1981

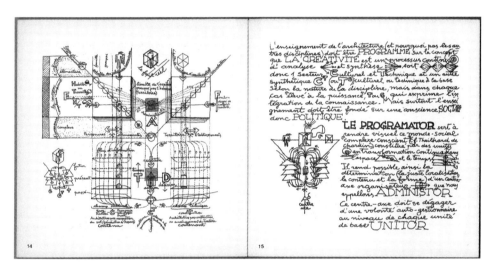

2.17 Justino Serralta, "Le programator," *L'Unitor*, 1981

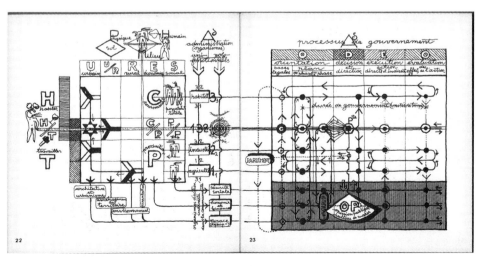

2.18 Justino Serralta, "Processus de gouvernement," *L'Unitor*, 1981

Meanwhile, Serralta oriented himself toward the search for a higher didactic order, derived from the *Modulor*. Le Corbusier himself provided the testimony:

> On March 30, 1953, Justino Serralta . . . having returned to Montevideo, wrote: "At the Architecture Department, I am adjunct professor of architectural projects; in my class, where teamwork is a basic, it's mandatory to use the Modulor and the [CIAM] Grid."[42]

From this first assignment, Serralta then passed to the elaboration of his own tools: *Unitor, Comunitor, Administor*, and so on. In the anxious years prior to the 1973 military coup, he labored in his atelier on a system that included urban structures in a socio-economic scheme, spanning from the territorial to the domestic level; utilized as a political program, it represented an almost utopian assertion of architecture as a weapon of revolution.[43] As he wrote in 1965:

> But to synthesize my fifteen years of professional experience I say: *less art and more technique*. To form professionals in order to produce housing for the community. To form professional architects, public servants; in the first place, good administrators, who know that with regulations, ordinances, and laws, one still makes Architecture, and furthermore, that this is the foundation for a good Architecture. To form architects who are aware of the current situation, but fundamentally who are prepared to solve the problems that in the not-too-distant future the people will require of them, and that today in fact are being addressed to architects working in the public service. To form architects who are aware of the fact that their technical function within this system will fundamentally be educational, and that their "master work" and contribution will enable the structural changes that the new community will require.[44]

Nonetheless, by the time Le Corbusier died in 1965, Serralta had confronted and argued with the traditional left about the socialist project supposedly inherent in the ideas of the Swiss master.[45]

Eventually Serralta would reject composition in favor of the concept of "organization" (this is another paradox for one who drew the most perfect version of the *Modulor*, though typical of the 1960s), and abandon architectural form. From his French exile he continued to construct, up until his death in October 2011, a complex theoretical system derived from that *Modulor* which, beginning with Teilhard de Chardin, eventually included Joël de Rosnay, René Thom, Karl Marx, among others.

And, of course, Le Corbusier.

Notes

Translated from the Spanish by Nicholas Sibille, with Helen Gyger and Patricio del Real.

1   Jorge Francisco Liernur and Pablo Pschepiurca, *La red austral. Obras y proyectos de Le Corbusier y sus discípulos en la Argentina (1924–1965)* (Buenos Aires: Universidad Nacional de Quilmes, 2008). Ramón Gutierrez, ed., *Le Corbusier en el Río de la Plata* (Montevideo: CEDODAL, Facultad de Arquitectura, Universidad de la República, 2009).

2   Kenneth Frampton, *Le Corbusier* (London: Thames & Hudson, 2001), 88. See also Fernando Pérez Oyarzun, ed., *Le Corbusier y Sud América*. *Viajes y Proyectos* (Santiago de Chile: Pontificia Universidad Católica de Chile, 1991).

3   Mauricio Cravotto to Armando Acosta y Lara (Dean), "Informe," December 14, 1938, Centro de Documentación e Información del Instituto de Historia de la Facultad de Arquitectura, Universidad de la República, Uruguay (CDI-IHA).

4   "Gral. Arq. Don Alfredo Baldomir electo presidente de la República," *Arquitectura* 196 (1938): 5, 7, 9. Baldomir was president from 1938 to 1942.

5   Nicholas Fox Weber, *Le Corbusier: A Life* (New York: Alfred A. Knopf, 2008), 303.

6   Le Corbusier to his mother, November 9, 1929. Fondation Le Corbusier.

7   Gervasio y Álvaro Guillot Muñoz, "Le Corbusier en Montevideo," *La cruz del sur* 27 (February 1930): 4.

8   Ibid.

9   Ibid.

10 Rodolfo Amargós, "VI Comunicación," January 2, 1925, CDI-IHA.

11 Rosendo Quinteiro to the Consejo de la Facultad de Arquitectura, March 5, 1927, CDI-IHA.

12 Rosendo Quinteiro to the Consejo de la Facultad de Arquitectura, November 8, 1927, CDI-IHA.

13 Juan A. Scasso, "Una beca de perfeccionamiento docente," *Anales de la Universidad* 146 (1939): 213.

14 Carlos Gómez Gavazzo to the manager of the Banco de la República O. del Uruguay, Octavio Morató, March 17, 1933, CDI-IHA.

15 Willi Boesiger, ed., *Le Corbusier et Pierre Jeanneret: Œuvre complète 1929–1934*, 5th ed. (Basel: Birkhäuser, 2006): 174, 178.

16 Carlos Gómez Gavazzo to Leopoldo Carlos Agorio, Dean of the Facultad de Arquitectura, September 16, 1933, CDI-IHA.

17 Carlos Gómez Gavazzo, "Paris–1933," *CEDA* 6 (July 1934): 20–21. Emphasis in original.

18 Carlos Gómez Gavazzo, "El atelier de Le Corbusier," *CEDA* 6 (July 1934): 18.

19 Carlos Gómez Gavazzo, "Concurso de Oposición para proveer el cargo de Profesor Adjunto de Proyectos de Arquitectura 1º al 3er años," March 1943, CDI-IHA.

20 Carlos Gómez Gavazzo, "Concurso de ideas para la ordenación arquitectónica de la Avenida Agraciada," *Arquitectura* 188, no. 1 (1937): 11.

21 "1er. Premio del concurso de la Cámara N. de Comercio," *Arquitectura* 188, no. 1 (1937): 7.

22 "Homenaje al Arq. Carlos Gómez Gavazzo," *Arquitectura* 216 (n.d., ca. 1945): 8.

23 "Concursos," *Arquitectura* 217 (n.d., ca. 1947): 35.

24 Carlos Gómez Gavazzo, "Concurso de Anteproyectos para el Palacio Legislativo del Ecuador," *Arquitectura* 219 (November 1948): 30.

25 "Centro de Asistencia del Sindicato Médico del Uruguay," *Arquitectura* 220 (December 1949): 25.

26 Mauricio Cravotto, "Presencia del Arquitecto Richard Neutra en Montevideo," *CEDA* 16 (December 1945): 23.

27 F. H. Bereterbide, A. B. Blanco, M. Cravotto, J. A. Scasso, "Plan Regulador de la Ciudad de Mendoza, República Argentina," *Revista del Instituto de Urbanismo* 8 (1942–1943): 24.

28 Liernur and Pschepiurca, *La red austral,* 277.

29 Liernur and Pschepiurca, *La red austral,* 389. See also Antonio Lapunzina, "Amancio Williams y la casa Curutchet," *Revista* (Buenos Aires) 3, no. 8 (1996): 43; Emilio Nisivoccia, "Les uruguayens. Le Corbusier, la política y la arquitectura en los sesenta," *Espacio* (Montevideo) 2 (2005): 137.

30 Justino Serralta, "Encuesta: 1950–1965: 15 años de arquitectura en el Uruguay," *CEDA* 29 (December 1965).

31 Serralta, "Encuesta," 22. Apparently Serralta wanted to suggest that, finally, he was able to do something connected with the architecture, since up until that point he had been patiently drawing sanitary pipes.

32 Le Corbusier, *Modulor 2 (la parole est aux usagers): suite de "Le Modulor" "1948"* (Paris: Editions de l'Architecture d'aujourd'hui, Collection Ascoral, 1955), 43–57.

33 Justino Serralta, *L'Unitor* (Paris: Justino Serralta, 1981). See also the Uruguayan edition: *El Unitor 2* (Montevideo: Fundación de Cultura Universitaria and Trilce, 1995).

34 Le Corbusier, *Modulor 2,* 18.

35 Le Corbusier, *Modulor 2,* 52.

36 Mauricio Cravotto to Julio C. Bauzá (Acting Dean), March 2, 1953, CDI-IHA.

37 "Plan d'ensemble de la Cité Permanente d'habitation. Premier projet, exécution en pisé," in *Le Corbusier: Œuvre complète 1946–1952,* ed. Willy Boesiger (Zurich: Editions Girsberger, 1953), 31.

38 Riccardo Gulli, "La huella de la construcción tabicada en la arquitectura de Le Corbusier," in *Las bóvedas de Guastavino en América,* ed. Santiago Huerta (Madrid: Instituto Juan

de Herrera, Ministerio de Fomento, 2001), 73; Henry Vicente Garrido, ed., *Arquitecturas desplazadas. Arquitecturas del exilio español* (Madrid: Ministerio de Vivienda, 2007).

39 Willy Boesiger, ed., *Le Corbusier et son atelier rue de Sèvres 35: Œuvre complète 1952–1957* (Zurich: Editions Girsberger, 1957).

40 Juan Pablo Bonta, *Eladio Dieste* (Buenos Aires: Universidad de Buenos Aires, 1963).

41 Graciela Silvestri, "Una biografía uruguaiana," in *Eladio Dieste 1917–2000*, ed. Mercedes Daguerre (Milan: Electa, 2003), 18.

42 Le Corbusier, *Modulor 2*, 34.

43 See Serralta, *L'Unitor.*

44 Serralta, "Encuesta," 22. Emphasis in original.

45 Justino Serralta, "Le Corbusier: primer arquitecto de comunidades," *CEDA* 29 (December 1965); Ricardo Saxlund, "Le Corbusier: la muerte de un genio de la forma arquitectónica," *CEDA* 30 (October 1966): 50.

3

MASS CULTURE AT MID-CENTURY

Architecture under a "New Humanism"

Noemí Adagio

> Karl Marx has given the concept of progress a unique interpretation which, I think, is not without influence on architectural practice. It would be proper, perhaps, to call this influence subconscious. The arts, said Marx, are determined in each era by the means of production specific to each era; and he gave to mechanized industry a high rank among those successive heavens to which man ascends through economic ameliorations. Thus, our art advances toward that supreme excellence in which architecture will become a form of industrial design.
>
> (Joseph Hudnut, 1952)[1]

From Sartrean existentialism to Arendt's "human condition," philosophy and the social sciences at mid-century invoked the concept of humanism in diverse ways. The most optimistic accounts envisaged "a renaissance beyond the limits of nihilism, darkness, and desperation," a sort of "World Age" which was developing "in the midst of the spiritual and moral ruins of the East and the West."[2] Taking into consideration their role in constructing the built environment, architects and urban planners turned their attention toward popular culture—especially its mechanisms for constructing identity and belonging—in the search for new meanings for architecture.

The responses were varied: Walter Gropius lamented the neglect of "the human element" in earlier decades; Richard Neutra defined architecture as the "human factor" and with an almost mystical attitude proposed making the most of available technologies in designing the house; some, aiming to find a higher purpose for their work, refused to limit themselves to solving only problems of comfort. Others, following Sigfried Giedion, aware of their place within their own era, hoped to understand it, as well as its meaning and direction, "from the inside"—and also from inside their own discipline, an important distinction.[3] Accepting the "tradition of the new" and also the need to broaden its horizons, the question of teaching modern architecture brought about a crisis in some of its values, reviving issues and concepts that had fallen into disuse, and returning to questions that had seemed already resolved. Now that spatial and formal explorations had produced radical departures in the conception and production of architecture, as well as the cultural significance of architectural practice itself, the central problem of planning a "modern" architectural education—at the moment of its institutionalization—was where to begin. Was the Bauhaus pedagogy the only possible solution? Could the relationship between art and technics operate outside the Germany of Weimar and Dessau? Was there any sense in restoring the "History" that had been roundly rejected by avant-garde groups?

Two figures of similar prominence and striking parallels, both convinced that the Rationalist approach of contemporary architecture was no longer sufficient, articulated post-Functionalist and post-Rationalist critiques while distancing themselves from those who pursued form as an objective in itself. Nevertheless, this was not the "postmodernism" elaborated through the historiography of critiques of the Modern Movement. Rather, I shall designate them "other modernisms."

Joseph Hudnut (1888–1968) was one of the first modernists in the United States, breaking with his Beaux-Arts training. Enrico Tedeschi (1910–1978), without receiving a modernist education per se, was a modernist by conviction, educated in and committed to the Italian intellectual circle around *Metron*; at the margin of the avant-gardes, and with less dogmatism, he managed to define his own ideas, while somewhat on the periphery of the international scene. Considering these two figures—who both believed that the "new tradition" defended by Giedion

3.01 *Metron*, October 1945

still had a future—in relation to one another, bearing in mind their specific backgrounds, presents us with a problem that is undoubtedly temporal rather than geographical, cultural, or generational in nature. In the late 1960s, the growing, and repeated calls for a design methodology that would eliminate any anxiety or uncertainty led to a schism between design-related concerns and architectural practice as a means of creating the built environment.

These two figures, standing as paradigmatic as regards their institutional roles (deans of Harvard and Mendoza, respectively) and their remarkable backgrounds (at Columbia, and at Rome and Tucumán), coincided in their defense of humanist values and their objection to the fascination of some architects with consumerist mass society. They were not detractors of modern architecture; without questioning its development or lingering over reproaches of its shortcomings, their positions presented a critical vision shaded by the postwar climate, which they experienced in different ways and which also distinguished them. Although it appears that they did not know each other personally, in 1951 Tedeschi presented Hudnut as "the brilliant exponent of a particular American critical humanist outlook," and as "progressive, anti-academic, with a Latin-flavored sense of irony."[4]

Marked by a "new humanism" that became necessary for both of them as they established themselves in architectural education, Hudnut and Tedeschi focused on "humanity" as the protagonist of history, but this viewpoint was not confined to populism, with all its ambiguities, or to the consumerism underlying popular culture and domesticity. On the contrary, they tried to restore the discipline's artistic engagement from within, and defended design as a means of "historical awareness" as well as a tool for the critical construction of culture. They both granted to history an important role in the education of the architect, and scrutinized the fallacies of contemporary culture. Hudnut was concerned about the fact that architects had adopted paradigms from other fields of production without even being aware of it, and Tedeschi enumerated the dangers that could waylay the architect as a result: the path of the "engineer," or the "architect of decorative composition and drawing," or the facileness of the "new academy."[5] To a large extent, both were unconvinced of the value of prevailing institutions in constructing an architectural culture and of the books that were supposedly educating architects.

Joseph Hudnut's "Other Modernism"

> The promises of a technological Utopia may kindle the mind of an architect but never the heart of mankind. Whatever may be our faith in industry and the machine, however firm our conviction of new freedoms and new horizons, we shall never satisfy with these the hunger of men for some assurance of beauty in their present lives and their present environment.
>
> It is my wish not to defeat modern architecture or stay its triumphant advance, but to exalt modern architecture by bringing it within the channel of a greater tradition.[6]

Educated in the Beaux-Arts tradition, Hudnut was among the first to install and defend modernism "institutionally" within the United States—even before Hitchcock and Johnson's contribution. From 1926, as dean of Columbia's architecture school, Hudnut radically changed the curriculum, placing history at the center of education, since it provided an "experience of architecture," teaching students to conceive it as a whole and granting them immediate access to its emotional content. Although his role in the modernization of education was recognized within the United States (reforms that were not accidental, but completely in tune with the country's euphoric calls for modernization), Hudnut would only gain international recognition after assuming the deanship of the Graduate School of Design (GSD) at Harvard in 1936, and appointing Walter Gropius as head of the architecture department the following year.

For some long time, Gropius had attracted attention as leader of the Bauhaus, and he continued to do so at the "Harvard Bauhaus" (as some called it); however, recent research gives credit to the job done by its dean over the seventeen years of his tenure, and maintains that the school would have been very different if Hudnut had not succeeded in implementing his ideas.[7] The disagreements between Hudnut and Gropius are useful in illustrating Hudnut's "other modernism." As is well known, Gropius maintained that in order to integrate art and architecture into modern life, the new architectural language should be "objectively valid" or derive from "visible scientific facts"—biological, physical, and psychological—and not from "emotion," which had driven the artist to a "sad isolation."[8] Furthermore, in contrast to Mies van der Rohe, who insisted that he did not pursue form for its own sake, Hudnut argued:

> I do not mean that form *is* beauty—the idea which colored the Renaissance— but rather that form is beautiful when it is also an expression of feeling . . .
>
> An architect models his building—subject to a thousand tyrannies of use, technical compulsions, costs, conditions of the site, and the vagaries of clients —as a sculptor models clay. He assembles, shapes, and defines volumes and masses; establishes their relationships to each other and to the whole; adds or takes away from each; emphasizes or suppresses, simplifies, elaborates, distorts. So far as an architect strives for expression he strives for form.[9]

Hudnut was still at Harvard when in 1953 he published *The Three Lamps of Modern Architecture* (identified as progress, nature, and democracy), condemning concepts emerging from fields external to architecture that had become the "measure of excellence in design," and whose introduction into architectural criticism was leading to confusion and mistakes. As he had done previously,[10] he defined architecture as "expression" and with a renewed conviction, explained that "the range of expression—or, let us say, the range of ideas to be expressed— is thus limited in each art." Hudnut added that: "the vehicle of expression is always plastic."

The fact that some architects had adopted the paradigm of progress as integral to the discipline led to problems concerning "representation" and expression: a high-tech building could represent modernity, but not necessarily express it. These subtle differences that Hudnut identified between the representation and the expression of technology clearly demonstrate a fear concerning what architectural practice could become, an anxiety in the face of the dangers (imagined or not) posed by architecture becoming a branch of industrial design.

It is important to emphasize that theories about form radically displacing the role of language were crucial for Hudnut, who believed that the achievements of the new architecture—"its severity of plane and contour, its precision, its devotion to fact"—did not arise, as many supposed, from advances in science, but from defeat, from "the failure of our architects to make use of new technological forms as the materials of 'artistic form'":

> Our materials, our technological devices, our rational (or at least our reasoned) preferences in character and symbol can be so reshaped and rearranged—

and without violence to the logic peculiar to each—as to become elements of form: of a form, concerned not with academic law and precedent, but with the free translation of idea and feeling beyond that possible to structures shaped by necessity.[11]

Hudnut did not conceive the building as an isolated object, not even in ideal terms, due to the influence of Werner Hegemann's concept of the integration of urban civic space, which Hudnut never abandoned.[12] Hudnut's response was focused on the technological experimentation that had taken architecture by storm. Let us recall some of the buildings under construction at this time: Mies's Crown Hall (1950–1956), the Saarinens' General Motors Technical Center (1948–1956), while Neutra had already built the Kaufmann Desert House (1946), and the Eameses were working on their Case Study Houses, to mention a narrow yet diverse overview.

Enrico Tedeschi's "Other Modernism"

[T]he technicality, for instance, which in the work of many architects has gone beyond the value of a mere practical fact to seek the role of being an expression of architecture, or even a fetish, through which the building "must" express its technical nature in order to be truly modern; the mechanization, which leads to giving buildings forms derived from machines, even if there is no relationship between the one and the other, in order to respond to another idol of contemporary mass society. . . . Then there is also quantification, in the naïve desire to amaze the public with the enormous dimensions, complex facilities, and large-scale investments that seem to be the principal requirement for some contemporary buildings, the pride of their authors and their owners.[13]

Enrico Tedeschi, along with Gino Calcaprina and Ernesto Rogers, was part of an ambitious project to develop Argentina's university system. Around two hundred specialists in different disciplines, most of them Europeans, were persuaded to settle in Tucumán (the smallest of the provinces) in order to foster the development of northern Argentina through a system of academic institutes distributed across the region, whose function would be to advise producers and government organizations. Forty new institutes in just three schools, and more than a hundred in total, demonstrate the ambition and the technological enthusiasm of the time.

At the end of 1947, Tedeschi, Calcaprina, and Rogers joined the Instituto de Arquitectura y Urbanismo (IAU), which had been founded by Jorge Vivanco a year before with the objective of working in relation to the environment, thereby radically modifying the curriculum that the Tucumán school had been following since 1939.[14] The pedagogy of the IAU was based on the needs of the region (which historically had been neglected) and aimed to revolutionize its architectural and urban development; as a consequence, seminars and practical workshops replaced lectures. The most important experiment in this context was the design and initial construction of a campus of colossal proportions conceived on the regional scale, made possible due to Perón's five-fold increase of its budget. Vivanco, along with Eduardo Sacriste, Horacio Caminos, and others at the IAU, were all former members of the Buenos Aires-based Grupo Austral, which had ceased functioning around 1941; nonetheless, this project can be seen in part as realizing the aims outlined in their 1939 manifesto.

At the IAU, Tedeschi joined an anti-academic philosophy, taking charge of teaching architectural history; his approach to and conceptualization of the subject are recorded in his *Una introducción a la historia de la Arquitectura*, published in 1951. His emphasis on the role of history in architectural education coincided with Giedion's call to awaken in students the ability to act as coordinator, to fuse in a work of art the elements contributed by other specialists.[15]

The enormous campus at Tucumán could not sustain itself over time, and remained as enormous, abandoned concrete structures. In early 1950, the IAU was dissolved altogether following pressure from the government to disseminate and promote Peronist doctrine.[16] Even through Tedeschi found himself obliged to stay there, he began to make connections with the National University of Córdoba, gradually establishing himself there in 1958. Whether predestined to settle in different regions of Argentina or motivated by the challenges that it posed, Tedeschi subsequently moved to the city of Mendoza to found the School of Architecture, the first private, secular institution, organized on the basis of a close relationship to the regional environment. At that moment, as dean of the new school, he had already spent twelve years in Argentina and had his own ideas about Argentinean society and university life.

I would like to focus a little on his paradigmatic work: the Facultad de Arquitectura de la Universidad de Mendoza (FAUM) of 1960. With a rich ambiguity—between architecture and engineering, between unity and modularity, between uniqueness and seriality—program and technical rationality are resolved in architectural form. I will not linger over the details, which have already been analyzed at great length;[17] rather, following Hudnut's instructions for the criticism of architecture that deserves the name, I will focus on the qualities of form, scale, and space, beyond the practical necessities fulfilled by the building.[18]

Resulting from "teamwork" (as proposed by Gropius and Giedion), not only was it realized with the engineers Roberto Azzoni (general calculation) and Diego Franciosi (design and assembly of prefabricated and prestressed elements), but also the program had been decided with the other professors in order to determine the needs of the school. Designer and commissioning client all at once, Tedeschi was a dean capable of managing and producing a project alongside his colleagues, in a horizontal manner characteristic of those who lead with authority.

3.02 Enrico Tedeschi et al., main facade of FAUM, 1960

3.03 Enrico Tedeschi et al.,
back facade of FAUM, 1960

Nothing in this building recalls Gropius, Le Corbusier, or Mies, because Tedeschi—
following his conception of the architect as an "inventor of architecture, not of
useless things"—even invented a technology appropriate to its economic
constraints and rapid execution, arriving at a solution based on prefabricated
elements (which, paradoxically, were used only once) as the response to a spatially
open program. An external structure at the scale of the whole building is
interrelated with a perpendicular structure at the level of each story, forming four
free floors for the workshops, studios, classrooms, and other spaces.

Despite the repeated elements (the V-shaped braces, the intermediate prestressed
beams), the idea of a whole prevails, since it is impossible to separate the parts
without jeopardizing the relationship between them (whether vertically or

horizontally). The building is undoubtedly a unit; it is a specific, particular response and in this sense, we may say "monumental." Tedeschi works at three scales: that of the building's structure, that of the everyday functional use of the building, and that of the urban scale of the building—set back from the street in order to create a meeting place—which provides the university with its social dimension. The building has no compositional hierarchy, not even to signal the entrance, which is displaced from the geometric center of the facade, and extends toward the public plaza in front of the building with the intention of incorporating this exterior space.

The product of conceptualization and previous modeling (calculus, knowledge of the behavior of materials, and foreseeing the form), the production process is very different from the resulting building. The observer is provided with clues to how the structure functions, but since the load-bearing system is unconventional, only an expert is able to comprehend it. With many of its parts prefabricated, the execution was rapid, but the assembly was painstakingly artisanal. The architectural work is diluted in the perfection of the pieces fabricated in workshops (with fastidious execution and expression) and is undoubtedly the result of a complex creative process. Conceptualizing and constructing is an aesthetic process; a complex compendium of verifications and decisions of various kinds that Tedeschi tried to synthesize in his *Teoría de la Arquitectura*—a theory of design, without doubt—which he wrote because "he did not know what reference to offer his students," as had happened a decade before with his *Una introducción a la historia de la Arquitectura*.[19] For Tedeschi, artistic value emerged from many different aspects that were impossible to enumerate or to rank, since deciding on the siting of the building on the lot was no more important than selecting a diagonal structure, or ensuring the optical control of the block. Tedeschi sought and achieved a suitable language for his own commitment to the notion of creation, but this was not a language "locked within itself"; in the Mendoza building, his architectural philosophy was implicit—Tedeschi emphasized that criticism and history could not be merely negative.

To determine the value of the structure in architectural terms, it is useful to compare Tedeschi's Mendoza building with contemporary experiments such as Affonso Reidy's Museu de Arte Moderna in Rio de Janeiro (1953–1968) and Clorindo Testa and SEPRA's Banco de Londres in Buenos Aires (1960–1966). All three buildings aimed to leave the maximum amount of free space possible— regardless of the fact that in the Banco de Londres this was later filled with various objects. In the bank, the structure—of undeniable expressive form—is intentionally sculpted, with voids of varied depths in which the repeating sections are lost in eye-catching dramatic plasticity. The expressively perforated cement is the antithesis of the minimal elements Tedeschi and Franciosi had defined. In the three buildings (architecture school, museum, and bank), the final exterior enclosure functions as screen and as solar filter. Their main staircases also fulfill active, well-defined roles: plastic and synthetic in the museum; overdramatized in the bank, since it is competing with many other objects; simple and minimal (in the style of Arne Jacobsen) at Mendoza. To sum up, a structure with trays hanging in the air for the museum and for the bank, and for the school a traditional

3.05 Enrico Tedeschi et al., main entry stair of FAUM, 1960

structure—apart from its great innovation—defined by the play of the membrane-like screen in sympathy with the perpendicular structure of the different levels.

We cannot call this work "regionalist" even though the facade functions perfectly as a climatic control and overall the building solves "specific regional problems" as discussed by Giedion.[20] Nonetheless, around this time Tedeschi recognized that the debates between regionalism and internationalism presented a certain potential: the possibility of in some way reversing the harmful effects of mass society, which—through homogenization and the demand for permanent novelty—distorted the meaning of architectural practice. Convinced of the need to "maintain architecture in its role as creative activity, and as such, as a constructive element of a culture," only in this sense—and never as a "stylistic option"—could the diversity brought by regional conditions and cultures be positive:

> Diversification seems as necessary in architecture as in mass society itself. . . . Because, if mass society represents a limit and a precondition for architecture as a practical activity, in its turn architecture can be an active precondition in order that mass society reaches—exceeding its current limits—a cultural situation defined within the history of humanity.[21]

From Harvard to Mendoza: Mass Culture Refracted at the Continent's Edges

In Tedeschi's critique of technicality as "fetish," of mechanization as the "idol of society," and of the complicated and lofty ostentations of technological form, we can see clear parallels with Hudnut. Despite the different philosophical and intellectual referents that formed them, Hudnut and Tedeschi shared key positions regarding architectural education. Neither believed in establishing fixed programs. The GSD program was constantly being revised—its objectives never definitively defined—just "as the conditions governing the professions in the world change"; in essence, Hudnut defined education as "life and preparation for life," an organic creative process.[22] Tedeschi also maintained that it was unnecessary to agree on curricula and programs (between large schools and small ones with few resources), that would never in themselves guarantee a good education; meanwhile he insisted that the student should acquire a "critical historical method" and that the "view of the problem as a whole" should be clearly articulated in order to establish the sense of a learning process conceived as a constant exercise throughout one's professional life.[23] Both considered that technical and professional education was not enough; as a consequence, both demanded a "humanist education"—the necessary connection between people and the environment in which they live and work. Hudnut sought a "third humanism," hoping that his students would distrust ready-made interpretations and "follow Dewey before the Air Force Academy," as he put it.[24]

To locate architecture as a discipline within the humanities, displaced from its technical definition, is to locate it strategically between the concreteness of society and the abstraction of the cultural framework. In this sense, history assumes a fundamental role, especially "the architectural work in history" which reveals "a problem of relations" and a "critical order" from which one can learn and gain experience about architecture. For Hudnut, the architectural work was a "document testifying to the invincible spirit" of humanity, that with time adopts a "human relevance" since it ceases to refer to pedestrian issues (such as construction, necessity, technique) and allows an understanding of more important issues, such as spatial solutions. The history that they championed was not an autonomous history (neither philology nor pure erudition) but one which—from the anxieties of the present—was operative in learning architecture, never just design; operative in recognizing the integrity of artistic (and therefore cultural) discipline. This much was not new: in all universities, history was changing from its traditional objective of narrating the temporal connections between events in order to explain specific phenomena, and beginning to be a history of architecture related by architects themselves.[25]

Since Tedeschi understood the discipline as an ongoing artistic activity that produced works of some interest, following the traces of major works that created—disseminated and vulgarized—the modes of expression, it was important for him to analyze and understand the "plastic, scale, and spatial" values, through the notion of "taste" as defined by Lionello Venturi.[26] This fanatical defense of the architectural work seems to have resulted from a presentiment of its disappearance with the emergence of notions of the "text" and the "death of the author"—a well-founded fear, as would be demonstrated a few years later.

The importance attached to the "work" is consonant with the emphasis given to "subjects" capable of imagining their work and its consequences: being responsible for their actions and interpreting society critically, controlling the contingencies that are supposed to determine scientific reality. The subjectivity that they defend is not an anxiety in the face of the infinite possibilities offered by the modern world; on the contrary, it is conceived as a positive—even inalienable—right. Undoubtedly, this stance is a response to the technological utopias then being developed, a reaction to the centrality of scientific develop-ment and its "paradigms of good sense." Still, Hudnut and Tedeschi were not opposed to technology; rather, they aspired to interpretative freedom—supposedly without constraints. For Tedeschi, this struggle for freedom, tinged with idealism, acquired the status of a political position in the context of contemporary Argentina as well as his previous experiences in Italy: the importance of the architectural work as "free expression of the artist's creative personality" had been confirmed by his experience with authoritarian governments, which had "led to failure and not to art, in the past as well as more recently in Germany, Italy, and Russia."[27] Even though mass society should not influence art, it did have an influence in the realm of "taste"; added to the harmful cult of the "isms,"[28] it pushed architecture—disposed to produce "works of permanent value"—toward the production of "objects of consumption." According to Tedeschi, some architects:

> Not only accept total freedom and the eclecticisms that derive from it, but also give the impression of continually looking for something newer, more sensational, more surprising, to jolt the public's indifference; exactly the reverse of what great artists do, who seem to elaborate the same idea throughout their whole lives . . . superficial and ridiculous fabrications are appearing, such as Marcel Breuer's trapezoidal windows, Niemeyer's campanile-toy at Pampulha, Edward Stone's folkloric and archeological pastiches, Philip Johnson's sophisticated academicisms . . . an incredible profusion of unformed and deformed forms, of diseased comic-book imaginings, all driven by a shameless and irresponsible exhibitionism.[29]

In the postwar climate of the United States, modern architecture was of necessity presented as functional (as Anderson has pointed out),[30] a situation that triggered Hudnut's reaction. In Argentina, the situation after Peron's downfall in 1955, and its consequences for the education system, clarified for Tedeschi the problem of the "college professor," lacking prestige or true autonomy (since he is dependent on government organizations). Only professors by vocation, devoted to study and research, would reverse the prevailing process, characterized by those "who teach design and composition based on their professional experience, aiming to make the student repeat their own certainties." In Argentina, there are no problems with the production of buildings, but rather an evident problem as regards culture: "a conformist, dogmatic mindset." While Hudnut, heir to the liberal intellectual tradition, rejected the overspecialization of the academic and professional fields and tried to mend the breach between theory and the reality of social life, Tedeschi assumed the consequences of professionalization as a fact that needed to be transformed—a distinction explained by their generational difference.

Differentiating and distinguishing the ideas of Hudnut and Tedeschi—a task impossible to avoid—is not the main objective of this chapter. Rather, it is to raise a more fruitful issue, which becomes evident through their convictions: the difference between architecture and design, which illuminates the crisis and the "non-crisis" of modern architecture from a perspective that is, at the very least, unusual. By now I hope to have demonstrated that these figures are not "postmodernists," despite the fact that Hudnut published "The Post-Modern House" in 1945.[31] If, by default, they are part of the "age of modernism" (to borrow an expression), they belong to it in two ways: because of their beliefs and because of the break that would soon take place—in the philosophical and sociological fields in general (phenomenology, structuralism, and semiotics) and in the architectural field in particular—which they did not support in any way.[32]

Discussions about architectural education took as their starting point the inadequacy of intuitive methods in relation to the complexity of the problems to be solved through design; these discussions also addressed the need to somehow fill the void left by the Functionalist approach to the shaping of form, which had insisted on the process of analytical induction as the basis of its design methodology. During the 1950s, this was an ever-present concern: not only in academic but also in professional spheres, the architect's social responsibility was discussed, and even "the client" emerged as a notion that should connect disciplinary knowledge to design practice. The most contradictory example is probably Richard Neutra, so sensitive toward the particularities of the client (via biology and psychology), while his projects undoubtedly emerged from his own artistic agenda. In this case, the "responsibility toward the client" is considered as being opposed to working on form itself.[33]

To think about learning rather than education makes clear the gap between "design," a social response that is compelled to work with the ever-changing resources of the market, which must be absorbed outside of architecture school, and "architecture," as a critical interpretation of culture, with a universe of self-referential resources belonging to the discipline. Architecture and design seem doomed to be strangers to one another: if there are directions (certainties) for design, it goes against the possibility of an individual search for answers. The solution to this tension is not to be found in a theory, nor in theories, plural, as the methodologies of the mid-1960s—beginning with Christopher Alexander—would prove.[34]

In the difficulty—or impossibility—that Hudnut and Tedeschi sensed at the moment of systematizing and planning a coherent (and definitive) education, would be rooted not the crisis of modern architecture, but its greatest achievement: a kind of assured vitality—a life of its own without a formula, as in life itself, but with no certainties on the outcome of the wager to continue the search for meaning, refusing to circumscribe architecture as merely a social practice. A system of representation based on plastic form (such as architecture) must presuppose the acquisition of certain given systems, but the results—far

METRON

CONSIGLIO DIRETTIVO

PIERO BOTTONI · CINO CALCAPRINA · LUIGI FIGINI · EUGENIO GENTILI · ENRICO PERESSUTTI · LUIGI
PICCINATO · SILVIO RADICONCINI · MARIO RIDOLFI · ENRICO TEDESCHI

DIREZIONE

LUIGI PICCINATO : URBANISTICA · MARIO RIDOLFI : ARCHITETTURA

SEGRETARIO DI REDAZIONE : MARGHERITA ROESLER FRANZ

SOMMARIO

3

EDITRICE SANDRON · ROMA · VIA MAZZAMURELLI N. 13

OTTOBRE 1945

3.06 *Metron*, October 1945, which included a translation of Hudnut's "The Post-Modern House"

from being determined by the process—are only formalized by the intentions that guide the designer.

Twenty years separated the histories produced by Giedion (1939) and Banham (1960),[35] two decades that were crucial in the search for meaning in the relationship between technics and society. At one extreme, the validity of the results—not to say the program—of the Bauhaus headed by Gropius; and at the other, Tedeschi's position, differentiating and distinguishing "architecture" from "design" and refusing to concede the determination of value to mass culture and the market. The disagreement is substantial: there is no possible exchange between architectural form and the applied arts—or, more properly, industrial design.[36] Let us recall that, in a similar fashion, Hudnut's words opening this chapter had pointed to the ghost of the superstructure determining everything, leaving no room for imagination or creation, leaving no room for art.

Notes

Translated from the Spanish by Helen Gyger, with Patricio del Real.

1   Joseph Hudnut, *The Three Lamps of Modern Architecture* (Ann Arbor: University of Michigan Press, 1952), 15. A translation of Hudnut's book was published in Argentina as *Las tres lámparas de la arquitectura moderna* (Buenos Aires: Nueva Visión, 1954).

2   Ruth Nanda Ansen, "Perspectivas del mundo," foreword to Walter Gropius, *Alcances de la arquitectura integral* (Buenos Aires: La Isla, 1956), 14. Translation of *The Scope of Total Architecture* (New York: Harper and Brothers, 1955). The Spanish version of the foreword contains several additional paragraphs (including this one) that do not appear in the English version.

3   "There are two different viewpoints for contemplating our own era: one that studies it from the outside, and the other that, being aware of forming part of it, hopes to understand its meaning and direction from the inside." Preface to the Spanish translation of Sigfried Giedion's *Architektur und Gemeinschaft* (Hamburg: Rowohlt Verlag, 1956): Giedion, *Arquitectura y Comunidad* (Buenos Aires: Nueva Visión, 1957), 7. This preface does not appear in the English translation, published a year later as *Architecture, You and Me; The Diary of a Development* (Cambridge: Harvard University Press, 1958).

4   Enrico Tedeschi, *Una introducción a la historia de la Arquitectura. Notas para una cultura arquitectónica* (Tucumán: Universidad Nacional de Tucumán, 1951), 130.

5   This idea of the facile nature of modern architecture had emerged in the local architectural culture some decades previously as a contemptuous response toward the unornamented modernist mass, and as a kind of resistance against modernity.

6   Hudnut, *The Three Lamps*, 11–12, 57.

7   Jill Pearlman, "Joseph Hudnut's Other Modernism at the 'Harvard Bauhaus,'" *Journal of the Society of Architectural Historians* 56, no. 2 (December 1997): 452–477.

8   With equal vehemence, Hudnut criticized the Bauhaus for its duality of art and craft, and he always opposed Gropius's efforts to implement a division between technical and art studies at the GSD. They also clashed over the history course: for Gropius, the past was in no way related to the present. Pearlman, "Hudnut's Other Modernism."

9   Hudnut, *The Three Lamps*, 9.

10 Especially in Joseph Hudnut, *Architecture and the Spirit of Man* (Cambridge: Harvard University Press, 1949).

11 Hudnut, *The Three Lamps*, 57.

12 Hudnut worked with Werner Hegemann for four years, until 1921, when Hegemann left the United States for Germany. Hudnut always bore in mind Hegemann's fundamental belief that the city is "the basis for architecture."

13 Enrico Tedeschi, *La arquitectura en la sociedad de masas* (Buenos Aires: Ediciones 3, 1962), 28.

14 See Franco Marigliano, "El Instituto de Arquitectura y Urbanismo de Tucumán. Modelo arquitectónico del Estado y Movimiento Moderno en Argentina 1946–1955" (PhD diss., Universidad Politécnica de Madrid, 2003).

15 Sigfried Giedion, "The Need for Basic Reform in Architectural Education," in *Building for Modern Man: A Symposium*, ed. Thomas H. Creighton (New Jersey: Princeton University Press, 1949), 118–124.

16 See Marigliano, "El IAU de Tucumán."

17 Leonardo Codina, "La estructura como instrumento de una idea. Enrico Tedeschi y el proyecto de la Facultad de Arquitectura de la Universidad de Mendoza" (master's thesis, Pontificia Universidad Católica de Chile, Santiago, 2004).

18 See Hudnut, *The Three Lamps*.

19 Enrico Tedeschi, *Teoría de la Arquitectura* (Buenos Aires: Ediciones Nueva Visión, 1962). This work was re-issued (and enlarged and revised) five times in less than fifteen years, demonstrating its impact in Ibero-American writings on architecture.

20 Giedion, "The New Regionalism," *Architecture, You and Me*, 148. Giedion's views on regionalism first appeared in "The State of Contemporary Architecture," *Architectural Record* (January 1954): 132–137.

21 Tedeschi, *Sociedad de masas*, 32–33.

22 In a 1929 discussion, Hudnut explained that "education is not instruction but experience." Pearlman, "Hudnut's Other Modernism," 457.

23 Enrico Tedeschi, "Sobre los métodos de enseñanza de la arquitectura," *Nueva Visión* 9 (1957): 30–32.

24 "Our students come to us eager for certainty. They will take to themselves every instruction which is framed within an assured knowledge of absolute truth, sonorously pronounced. For that the religious leader has every advantage over the humanist. It is not from reason (as they imagine) that our students draw that obscurantism of ritual and vision upon which they build their sterile castles of steel and glass." This is a clear allusion to Mies. Joseph Hudnut, "Humanism and the Teaching of Architecture," *Journal of Architectural Education* 4 (Winter 1961): 12–16.

25 Regarding the US context, Anderson analyzes a similar movement he pinpoints as beginning in 1960. Stanford Anderson, "Architectural History in Schools of Architecture," *Journal of the Society of Architectural Historians* 58, no. 3 (September 1999): 282–290.

26 Tedeschi explained that Lionello Venturi's notion of "taste" demonstrated "the artist's preference for certain formal structures, accompanied by the demonstration of other preferences of a technical, or more generally human, character. These are the result of the choices that his taste has made among the different possibilities offered by the environment in which the artist lives, and that conditions his culture, his ideals, his moral or religious attitude; hence in this sense the artist is a man among men." Tedeschi, *Sociedad de masas*, 28.

27 Tedeschi, *Sociedad de masas*, 27.

28 "Functionalism, neoplasticism, constructivism, organicism . . . many of these labels have been used in architecture. Gathering in groups and giving themselves a name and publishing manifestos of their intentions—which in reality should be explicated by their architectural works—and somehow pretending to be the proprietors of a truth . . . is a procedure derived from a democratic type of political structure belonging to mass society . . . bringing confusion and mistakes." Tedeschi, *Sociedad de masas*, 29.

29 Tedeschi, *Sociedad de masas*, 30–31.

30 See Stanford Anderson, "The 'New Empiricism—Bay Region Axis': Kay Fisker and Postwar Debates on Functionalism, Regionalism, and Monumentality," *Journal of Architectural Education* 50, no. 3 (February 1997): 197–207.

31 Joseph Hudnut, "The Post-Modern House," *Architectural Record* 97 (May 1945): 70–75.

32 As stated by Anderson, the image of a naive modernism, a Functionalist Rationalism in an atmosphere of narrow-mindedness, was useful for postmodernists. Anderson places responsibility for this interpretation on various historians, and lucidly argues that neither in theory nor in practice was modern architecture conceived in a restrictive way. Stanford Anderson, "Types and Conventions in Time: Toward a History for the Duration and Change of Artifacts," *Perspecta* 18 (1982): 108–117.

33 See Louis Kahn, Paul Weiss, and Vincent Scully, "On the Responsibility of the Architect," *Perspecta* 2 (1953): 45–57.

34 Christopher Alexander, *Notes on the Synthesis of Form* (Cambridge: Harvard University Press, 1964) was a response to the need to systematize design pedagogy. At the opposite end of the spectrum and under the belief that theory kills creative ability, Christian Norberg-Schulz published *Intentions in Architecture* (Oslo: Universitetsforlaget, 1966).

35 Sigfried Giedion's Charles Eliot Norton Lectures for 1938–1939 were published as *Space, Time, and Architecture: The Growth of a New Tradition* (Cambridge: Harvard University Press, 1941); Reyner Banham, *Theory and Design in the First Machine Age* (London: The Architectural Press, 1960).

36 These differences point to the paradox that Tedeschi was asked to publish his writings in *Nueva Visión*, a magazine focused on aspects of design.

Part II

TECHNO-CULTURAL ASSEMBLAGES

4

PRE-COLUMBIAN SKINS, DEVELOPMENTALIST SOULS

The Architect as Politician

Luis Castañeda

Even when compared to the canons of some of its peers among the "emerging" fields of architectural history, the canon of twentieth-century Latin American architecture is limited in size, with the work of a very small number of figures active during the middle of the twentieth century—Luis Barragán (1902–1988), Oscar Niemeyer (b. 1907), or Carlos Raúl Villanueva (1900–1975)—still taking up most scholarly attention. Problematically, the literature has failed to give serious consideration to the political and social connections that lie behind the architectural success of these inevitably male, and almost inevitably elite, "form givers." Beyond the stories of these selected few, the literature has especially neglected the careers of many politically involved architects and planners in Latin America, even when many of these figures took part in fundamental episodes in the recent political history of the region.[1]

This chapter presents a comparative analysis of the careers of two such architects: Pedro Ramírez Vázquez (b. 1919), the architect with the most intimate ties to presidential power in twentieth-century Mexico; and Fernando Belaúnde Terry (1912–2002), trained as an architect in the United States during the 1930s, and elected president of Peru twice—first, between 1963 and 1968, until he was ousted from power by a military coup, and subsequently between 1980 and 1985, after Peru's decade-long military government ended. Ramírez's prolific career is relatively well known. Some of his works, such as his National Museum of Anthropology and History in Mexico City (1964), stand among the most visible examples of Latin American architecture. Yet, Ramírez's complex role as the architect of the official image of Mexico's single-party state, which involved more than the making of buildings, has yet to receive serious attention. Belaúnde, who built far fewer buildings than Ramírez, is primarily known for his rise and fall from presidential power, but the interrelation between political and architectural practices that makes his career of interest has not quite received its due.

This chapter does not attempt to provide a comprehensive account of the careers of Belaúnde and Ramírez, but instead examines a fundamental intersection between them: their politicized appropriations of the ancient American pasts and legacies of indigenous cultures of Mexico and Peru. A fundamental ingredient of the political culture of Latin America, this type of appropriation was especially relevant in Peru and Mexico, modern-day sites of the most significant early building cultures of the region. This appropriation was also a fundamental ingredient of the developmentalist agendas with which the careers of both Belaúnde and Ramírez were aligned. Indeed, Belaúnde and Ramírez understood architecture and regional planning as the primary tools for national economic emergence. They also understood architecture's cultural work as central to the identity-building efforts of "developing" countries.[2]

In the work of both Belaúnde and Ramírez, appropriations of the heritages of the ancient Americas were most interestingly featured in works not narrowly architectural in nature. These appropriations were perhaps best expressed in what we may describe as architecture's "expanded field"—the multiple intersections between architecture and other forms of visual, literary, and artistic production.[3] In the case of Ramírez, I will examine his role as the figurehead of a major propaganda event: the production of a network of museums of national culture, the centerpiece of which was the National Museum of Anthropology. Through an analysis of his virtually unknown political writings, I will inscribe Belaúnde's identity as an architectural and political *provocateur* within the ideological panorama of twentieth-century Latin America.

Parallel Lineages of Architecture and Power

Belaúnde and Ramírez met more than once, but one of their most significant encounters took place in 1964, as Ramírez visited Lima as part of a promotional tour for his newly completed Museum of Anthropology. Belaúnde was eager to build a similar facility in Lima, but the project for a new anthropology museum in Peru did not materialize until the late 1970s, after Belaúnde had been ousted from power for the first time.[4] Beyond this episode, at first glance the stories of these two architects could not seem more different.

Belaúnde was born into an elite household in Lima, one with significant ties to politics. Rafael Belaúnde (1886–1972), Belaúnde's father, was a diplomat who served as a member of several presidential cabinets. Pedro Diez Canseco (1815–1893), a great-grandfather of Belaúnde's and a military figure, served as interim Peruvian president during three episodes of political instability, in 1863, 1865, and 1868. Belaúnde's father was persecuted during President Augusto B. Leguía's second administration (1919–1930), a fact that precipitated the family's move to Paris in 1924, when the would-be architect was twelve years old. For the next decade, Belaúnde spent his adolescence and early youth in various cities in Europe and the United States. Rafael Belaúnde was a founding member of the University of Miami, where Fernando would begin his training as an architect. Belaúnde finished his degree at the recently established School of Architecture at the University of Texas at Austin, in 1935. As we shall see, this formative experience of exile defined Belaúnde's plans for a presidential agenda.

After a short stint working as an architect in Mexico City in 1936, Belaúnde returned to Lima the following year, and founded *El Arquitecto Peruano*, the country's first architectural magazine, and his political mouthpiece for the subsequent three decades.[5] Although he mostly designed single-family houses, it was through the promotion of social housing that Belaúnde became involved with municipal and, gradually, national politics. As Wiley Ludeña has shown, Belaúnde and his associates' interventions in the popular press fueled the planning debate, a debate that eventually led to the creation of a series of government agencies devoted to planning by the mid-1940s in Peru.[6] The mid-1940s also witnessed the beginning of Belaúnde's career as a politician. In 1945, Belaúnde was elected to Congress, serving in that capacity until President José Bustamante's administration—where Belaúnde's father served as a cabinet member—was cut short by the coup of military commander Manuel Odría in 1948. As a congressman, Belaúnde oversaw the creation of one of the earliest housing projects in Peru, the Unidad Vecinal No. 3, a set of neighborhood units built in the outskirts of Lima starting in 1945. Subsequently, Belaúnde was a key figure in providing political support for a series of regional planning projects for Peru: the city plans for Chimbote (1948) and Lima (1948) by the New York-based firm of Town Planning Associates (TPA), headed by Josep Lluís Sert and Paul Lester Wiener. Part of a

Una Gran Realización Peruana:
La Unidad Vecinal No. 3

Setiembre 1949
S/o. 4.50
★

El Arquitecto PERVANO

4.01 Luis Dorich et al.,
Unidad Vecinal No. 3, Lima,
1948

well-known sequence of plans for Latin American cities by TPA, neither of these two schemes was implemented due to a combination of shifting political tides and financial troubles, and despite Belaúnde's aggressive endorsement.[7] Belaúnde spent the better part of the 1950s as the dean of the School of Architecture at the Universidad Nacional de Ingeniería. He also founded his own political party, Acción Popular, in 1956, beginning the run for his first presidency later that same year.

Unlike Belaúnde, Ramírez Vázquez never held political office, and his last significant encounter with political power took the form of an unsuccessful run for the Mexican senate in 1989.[8] The irony of the episode is hard to miss: Ramírez never managed to become a senator, even when his firm had designed the building for the new Legislative Palace in Mexico City in 1981. Yet long before this episode of electoral failure, Ramírez's practice as an architect was defined by his close relationship with various state institutions in Mexico.

If Belaúnde was caught in the maelstrom of presidential politics more than once during his career, Ramírez had no such instability to contend with. The Partido Revolucionario Institucional (PRI) controlled the state that Ramírez worked within through the entirety of his career. Unlike Belaúnde, Ramírez was born into a family of humble origins and no real political ties, but shares his story of unlikely social prominence with two of his brothers, Manuel (Secretary of Labor 1948–1952) and Mariano (Justice for the Mexican Supreme Court 1947–1973).[9] The unlikely careers of these three members of the Ramírez clan would be hard to explain, if not seen in relation to the populist measures of the Mexican state. In particular, the early years of Ramírez's career were defined by efforts on the part of this state to democratize public education. A protegé of official intellectual Jaime Torres Bodet (1903–1974) since his days as a student at Mexico City's Escuela Preparatoria, Ramírez's entry into politics in Mexico began through his involvement with the construction of public schools between 1944 and 1947, during the presidency of Manuel Avila Camacho (1944–1950). School building had been a highly politicized field in Mexico since the early twentieth century: not only was it dominated by the patronage of the expanding single-party state, but architecture's relationship to public education was directly informed by ideological transformations in the highest spheres of power.

Ramírez's first major commission, the 1953 School of Medicine at Mexico's University City for the National University (UNAM) completed with Héctor Velázquez and Ramón Torres, was part of the largest state-sponsored building campaign devoted to education in modern Mexico.[10] In 1958, as a result of his ties to Torres Bodet, Ramírez was named head of the federal program for the construction of schools (CAPFCE), an agency tied to the Secretariat of Public Education, and occupied this position until 1964. Ramírez won his first international accolade, the grand prize at the 1960 Milan Triennale, for his design for the prototype for a hybrid between a house for teachers and a school intended for rural communities, a project he tested in various rural contexts in Mexico during his official appointment at CAPFCE. In conjunction with the National Institute of Anthropology and History (INAH), CAPFCE subsequently became the sponsor of Ramírez's Museum of Anthropology, and of a network of national museums of culture planned for Mexico of which it was to be the primary venue. And, as I will show below, it was also in the context of these works that Ramírez's appropriation of the Pre-Columbian past took on its clearest ideological contours.

Looting Architecture out of the Tombs

Most covers of *Time* magazine during the 1960s included portraits of important world figures, but its March 12, 1965 issue, which featured Belaúnde on its cover, must have surprised its readers. Belaúnde, presented as a "Latin American architect of hope," was the only Latin American architect, and one of few politicians from the region, to ever achieve this level of visibility in the United States. Yet, the article in *Time* presented Belaúnde as a perfect "Alianza President" —a leader who wholeheartedly embraced the interventionist, Kennedy-era "Alliance for Progress" initiatives—and hence as a strategic ally in the middle of unfriendly territories in the hemisphere.[11]

The title of the *Time* article—"The New Conquest"—was inspired by one of Belaúnde's many publications, a 1965 book titled *Peru's Own Conquest*, an expanded English edition of an earlier work published in Lima in 1959, in the heat of Belaúnde's second presidential campaign.[12] Compiling several short texts Belaúnde had published in *El Arquitecto Peruano*, *La Conquista* was far more than a campaign pamphlet. The book gathered Belaúnde's impressions as he traveled through much of Peru, and wove together proposals for new public works projects

4.02 "Peru's President
Belaúnde: A Latin American
Architect of Hope." Cover of
Time magazine, March 12,
1965

with romantic allusions to the monuments of the Incas. The most important new project proposed in *La Conquista* was the Forest-Edge Highway. The Highway was intended to cover the greater part of the Eastern stretch of the Andean mountain range, where the Peruvian section of the Amazon rainforest begins. An undertaking of continental scale that Belaúnde promoted heavily in his publications, the Highway would connect Peru with several of its neighboring countries while "colonizing" a large section of this *terra incognita* for agriculture and trade.[13]

The Highway stood in clear relationship to the operation fostered by Juscelino Kubitschek's administration in the construction of Brasília (1960), one intended to create both an emblematic new national capital and a hub of continental transportation and commerce. The Highway's similarities with US Depression-era projects, which Belaúnde had been introduced to as a student in 1930s America, were also clear.[14] The continuity between the Chimbote plan by TPA that Belaúnde championed aggressively during the 1940s, which intended to create a regional hub out of the coastal Peruvian city, and the Forest-Edge Highway, which was meant to implement continental-scale planning for Peru, is also hard to miss. Preliminary studies for the Highway were undertaken not by TPA, but by another US planning firm most often identified by its acronym: the Manhattan-based firm of Tippets, Abbets, McCarthy, Stratton, or TAMS, known for its work with the US Army Corps of Engineers since its wartime foundation in 1942.[15]

For all that was modern and "American" about Belaúnde's project, the maps and diagrams with which Belaúnde illustrated the project were also quite similar to images of the network of Inca roads he included in *La Conquista*. In one striking

ROAD TO NEW LANDS

A Proposal for South American Development

FERNANDO BELAUNDE TERRY

PROPOSED SOUTH AMERICAN FOREST-EDGE HIGHWAY

▨▨▨	High Forest
▩▩▩	Mountains
▬▬▬	Colonization Highway
╫╫╫╫	Railroads
------	Canal proposed by engineer Gabriel del Mazo

IF THEIR PLANS for economic development are to be successful, the South American countries must look at their problems from a regional, not a narrowly national, point of view. It is only just and prudent to think in terms of a South American highway program, transcending our national schemes.

For the system of highways and railroads plays two vital roles in the process of development of a continent: one is communication, and the other is colonization, permitting the incorporation of new cultivable lands, providing both a place for some of the ever-growing population to live in and increased food supplies. Only thus can the "geography of hunger" of which Josué de Castro wrote be resolved into a geography of abundance.

The highways man builds must complement rather than duplicate the means of travel and transportation provided by nature. Let us consider the pre-existing, natural communication routes in South America. In the first place, the entire land mass is surrounded by ocean. Since this assures a means of communication around the periphery, coastal roads, like the section of the Pan American Highway that runs along the Pacific coast, are not really those it is most urgent to build, although they have already cost several governments a great deal of money. There are other routes that should have priority. The great river systems of the Orinoco, the Amazon, and the Paraná form an unfinished network. Some time ago, the distinguished Argentine engineer Gabriel del Mazo proposed building canals to unite these waterways. Despite the topographical obstacles that would have to be overcome, it is an interesting suggestion. But this project, which would be carried out entirely within the lowland forest area, would provide communication without meeting the basic need for colonization of the areas that, because of healthful climate and soil fertility, are most suitable for settlement—and so urgently needed for the exploding population.

Both of these objectives would be attained by a road I have proposed, to be called "The South American Forest-Edge Highway," since, instead of going through the middle of the unhealthful, low-lying jungle, it would skirt the edge of that "green hell," taking a level part-

way upstream on the rivers. Here, because of the altitude, the land would offer climatic and ecological advantages despite the tropical latitudes.

For 2,500 miles, this highway would cross the eastern slopes of the Andes in Colombia, Ecuador, Peru, and Bolivia, linking points of navigation on the Meta, Amazon, Beni, and Paraná Rivers, and providing access to the great ocean ports at the Orinoco delta, at Belém, and on the River Plate estuary. It would have direct rail connections to the Pacific ports of Antofagasta and Arica, Chile, and Matarani, Peru, and only short extensions of present rail lines would be needed in Colombia and Ecuador to tie in it with ports there. Construction of such a road would make an additional 150,000 square miles of land available for colonization and cultivation. There is not a single South American country that would not benefit, directly or indirectly, from such a route.

The high forests of Bolivia, Peru, Ecuador, and Colombia all offer very similar conditions. Between the Amazon plateau and the snow-covered peaks of the cordillera there is a whole gamut of climates and soils suitable for a variety of crops. The forest resources of this region and the mineral and petroleum possibilities are well known. Diversified production can be achieved by locating the road so that it will serve areas that have been selected because their ecological conditions will assure the success of the desired crops or other enterprises. For every mile of road built, some 3,500 acres will be made accessible. Whatever the cost per route mile may be, the cost per acre of new land will be insignificant, far less than the selling price of the high forest acreage that is now accessible by road.

The pressing and extremely difficult problem of agrarian reform will not be solved or simplified unless there is a radical increase in usable land. It is not a simple question of the redistribution of an abundant commodity: something that is in short supply must be multiplied. It must be recognized that, although our continent has a very large area, we lack sufficient arable lands. Among others, the case of Peru is dramatic: here, because of the large desert areas and the lack of access to most of the rich lands, only 1.4 per cent of the territory is at

present under cultivation.

A project like this, that will benefit all the countries of South America, should recommend itself to the new Inter-American Development Bank as worthy of support. It is no secret that very often the South American countries have made separate and conflicting requests for credit, appearing in the role of competitors instead of acting harmoniously as associates concerned with a common objective. And, clearly, the mission of the Bank is to help us solve our problems on a continental scale.

Such a project as this one would be conducive to a spirit of union and fraternal cooperation among all our countries. There would not be the competition or the suspicions, the opposition that has sterilized or delayed every beneficial recommendation, in short, the negative diplomacy of paralysis that has been so harmful to our countries. It would elicit, on the contrary, the positive and fertile diplomacy of cordial joint action.

This is the fastest way to achieve the much desired geography of abundance. ❧

Revista de la O.E.A. publica proposición continental del Arqto. Belaúnde Terry.

4.03 Fernando Belaúnde Terry, "Road to New Lands," *El Arquitecto Peruano*, 1960, reprinted from Organization of American States publication

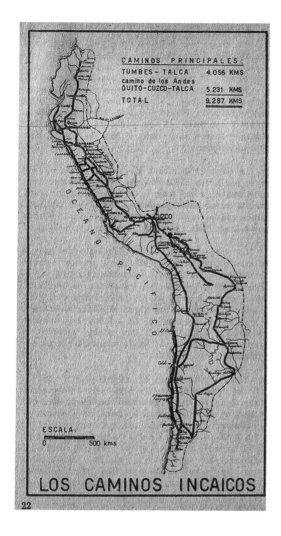

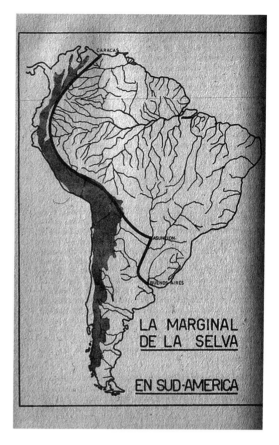

4.04 (above left) "The Inca Roads," in Fernando Belaúnde Terry, *La Conquista del Perú por los Peruanos*, 1959

4.05 (above right) "The Forest-Edge Highway in South America," in Fernando Belaúnde Terry, *La Conquista del Perú por los Peruanos*, 1959

case, he layered the schematic lines of the new highway over the shaded surface of the Andean mountain range on a map of South America, a visual arrangement that seemed to superimpose them not against the current geographic extension of Peru, but against the much larger territory occupied by the Inca empire until its downfall in the sixteenth century. Belaúnde's point was thus well made: the use of modern technologies of engineering and construction would bring back Peru's long-lost "Golden Age" while simultaneously inscribing the country within mid-twentieth-century spheres of commerce and geopolitical influence.

Belaúnde published one more travel book in 1960, *Pueblo por Pueblo*, a work that was even more explicitly tied to imperial desires vis-à-vis Latin America's territories than *La Conquista*.[16] Indeed, Belaúnde was by no means the first to publish a book of this kind. Traveling European natural scientists such as Alexander von Humboldt (1769–1859) and Clements Markham (1830–1916) had published extensive travelogues discussing Peru's geography and history during the nineteenth century.[17] Such US travelers as Ephraim George Squier (1821–1888) and Hiram Bingham (1875–1956)—most famous for his "discovery" of the site of Machu Picchu in 1911—were also members of this long lineage.[18] Squier, a US diplomat in Latin America between 1846 and 1869, published his *Peru: Incidents of Travel in the Land of the Incas* in New York in 1877, a book that is almost identical to Belaúnde's and may well have served as a model for the architect's publication.[19] Not only did Belaúnde and Squier record their passage through almost the exact same sequence of ancient Peruvian monuments both visually and in written vignettes, but also what they had to say about these monuments was strikingly similar. The resemblance is most obvious in their discussions of Inca sculpted rocks, objects to which they both devote considerable attention. Taking a close look at one such artifact, Squier found in the mysterious grooves of its surface an explanation of the Incas' sophisticated understanding of irrigation channels.[20] Writing about a similar monolith, the so-called Stone of Sayhuite, in Cuzco, a sculpted rock that included an intricate pattern of carvings resembling the model of a mountain city, Belaúnde pushed the point even further. "Standing before this monolith," he wrote, "we experience the same emotion that overcame us as we encountered Champollion's famous Rosetta stone at the British Museum, [an artifact] that shed so much light on the history of ancient Egypt." "Here, there are no two-dimensional hieroglyphs . . . but there is something even more precise: a three-dimensional version of a community in ancient Peru, both in its urban and rural aspects."[21]

Belaúnde's need to situate his encounter with the Stone in a museum-viewing context is significant, since it was in the context of European museums that he was first introduced to Pre-Hispanic artifacts during his early exile. But his allusions to Egypt as a point of reference indicate the extent to which his presidential

4.06 (below left) Stone of Sayhuite, in Fernando Belaúnde Terry, *Pueblo por Pueblo*, 1960

4.07 (below right) "The Rock of Concacha—Front and Back," in Ephraim George Squier, *Peru: Incidents of Travel and Exploration in the Land of the Incas*, 1877

Arriba, la piedra de Sayhuite, en forma de taza, con un diáme-tro de unos cuatro metros y una altura de más de dos. La his-toria se lee aquí en tres dimensiones. La flora, la fauna, el cul-tivo, el riego, el urbanismo y la arquitectura se mezclan para dar una completa visión del pasado. Morúa refiere que esta re-gión de Concacha estuvo dedicada a los hombres seleccionados para concebir las grandes construcciones y para asumir la-bores directivas. Era, evidentemente, un centro cultural.

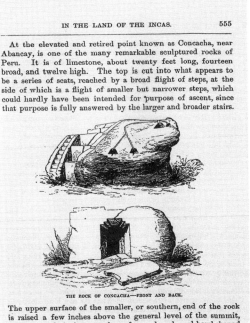

IN THE LAND OF THE INCAS.   555

At the elevated and retired point known as Concacha, near Abancay, is one of the many remarkable sculptured rocks of Peru. It is of limestone, about twenty feet long, fourteen broad, and twelve high. The top is cut into what appears to be a series of seats, reached by a broad flight of steps, at the side of which is a flight of smaller but narrower steps, which could hardly have been intended for purpose of ascent, since that purpose is fully answered by the larger and broader stairs.

THE ROCK OF CONCACHA—FRONT AND BACK.

The upper surface of the smaller, or southern, end of the rock is raised a few inches above the general level of the summit, and has sunk in it a number of round and oval bowl-shaped cavities, varying from four to nine inches in diameter, and from three to six inches deep. From one of these nearest the edge of the rock leads off a little canal, which conducts over the side of the rock, where it branches, leading into four reservoirs cut in the stone, in the style of pockets, the two larger ones capable

journeys in Peru were concerned not just with actual monuments but also with the numerous layers of meaning with which they had long been invested by "imperial" eyes. In her now-classic study of European travel writing, Mary Louise Pratt explores the imperial ambitions of eighteenth- and nineteenth-century travel narratives about the American continent. In her discussion of Humboldt's writings, Pratt argues that a model for the "archaeological rediscovery of America" was Egypt. As in the case of the New World, "[there] too," she claims, "Europeans were reconstructing a long history through, and as, 'rediscovered' monuments and ruins."[22] Belaúnde reworked this gesture, articulating a "rediscovery" of his own country in the centuries-old language of empire.

Primitivist readings of Peru's history not only defined Belaúnde's early discourse, but they would also never leave him. Asked about the future of his political party in 1987, after the end of his second presidential term, Belaúnde responded in the same terms he would have used to describe its agenda three decades earlier. "I have asked the young [members of my party], to devote themselves to the search for new and unsuspected aspects of the inexhaustible Andean legacy. I have asked them, yet again, to '*Loot the doctrine out of the tombs*'."[23] By the late 1980s, Belaúnde's statement was a bit outdated, but it was right at home in Peru earlier in the twentieth century. The idea that Peru's deep past could be mined as a source of cultural, aesthetic, and political doctrine was not only current among generations of European and US travelers, but Latin American and Peruvian intellectuals also subscribed to similar views. Intersecting with Belaúnde's ideas is at least one strand of radical thought: José Carlos Mariátegui's (1894–1930) notion of *indigenismo revolucionario*. Mariátegui's formulations included political and aesthetic dimensions and advocated avant-gardist modes of image-making and writing as the means to access and unlock the revolutionary energies of the Andes. Much of Mariátegui's intellectual production emerged from his work as a journalist and publisher, especially from his publication *Amauta*, founded in 1926.[24] Like Belaúnde's, Mariátegui's aesthetics emerged as the product of the experience of exile, specifically in relation to a journey to Europe and back, between 1920 and 1923.[25]

Belaúnde's understanding of the revolutionary energies of the Andes was ostensibly more conservative than Mariátegui's, but both figures produced similarly Primitivist readings of the Andean past and present as a repository for political transformation.[26] Much more closely than Mariátegui, however, Belaúnde followed the lead of conservative thinkers among his fellow Peruvian intellectuals, some of them members of his own family. Intellectuals like Víctor Andrés Belaúnde (1883–1966), the architect's uncle, theorized deeply conservative readings of the Andes as a site for political energies in such writings as *Peruanidad*, a book published in 1947.[27] Abstracting the present-day "Indians" from the picture, Belaúnde's *Peruanidad* construed as members of the same genealogy the grand Inca elites of old and his own aristocratic class.[28] Presaging the arguments of his architect nephew, Belaúnde's book also urged future rulers of the country to follow in the footsteps of their alleged predecessors, taming Peru's savage territory, building roads and dams like theirs, and replicating their fair, "collectivist" institutions.

This trajectory of appropriations of the Pre-Columbian past formed the basis of Belaúnde's pre-presidential claims. Nevertheless, the visual and written narrative of his journeys not only revealed his "imperial" ambitions for Peru; it also served more instrumental political purposes. As disseminated in his travel books and in *El Arquitecto Peruano* beginning in the mid-1950s, images of Belaúnde's admiration for the landscapes and monuments of Peru emphasized to potential voters that the architect was the first presidential candidate to visit Peru's remote locations,

4.08 "Dos Arquitectos en
Puno," *El Arquitecto
Peruano*, 1961

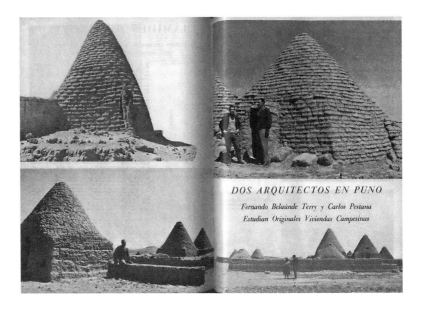

DOS ARQUITECTOS EN PUNO

*Fernando Belaúnde Terry y Carlos Pestana
Estudian Originales Viviendas Campesinas*

and presented him as a messianic unifier of the long-divorced social, economic, and ethnic groups of the country. Belaúnde's travels were thus acts of propaganda in the most obvious sense, acts with clear Latin American parallels, both past and contemporary. For instance, Belaúnde's travels were highly reminiscent of Fidel Castro's movements through the Cuban countryside and Sierra Maestra in the context of the 1959 revolution—a resemblance that *Time* magazine duly noted in 1965, while anxiously emphasizing to its readers that Belaúnde and Castro did not quite share the same political plans or attitudes toward the United States.[29] Belaúnde's national journeys even more clearly recalled the trips made by Mexican president Lázaro Cárdenas during his presidential campaign of 1934. Cárdenas, a former teacher turned aspiring president, visited much of rural Mexico on horseback and by car, eliciting widespread admiration at home and abroad through narratives in word and image, and eventually reached the highest political office.

Archaeologies of Power in Mexico

An archival photograph from mid-1964 documents a visit by a group of architects, politicians, and bureaucrats, to the brand-new National Museum of Anthropology and History in Mexico City. In the image, Ramírez Vázquez examines the display of the Aztec chamber, the most important room in the Museum, alongside Mexican president Adolfo López Mateos (1958–1964), other officials, and designers from his office. Although the photograph seems to show us the architect–politician as the figurehead of a presidential propaganda operation, the image is strangely off-kilter. Most noticeably, the spotlight in the room falls on the renowned Aztec Calendar Stone, to the right, while the visitors look intently in the opposite direction. Upon close examination, it becomes clear that this image is a montage where architects and politicians have been added *post facto* onto a staged photograph of the empty Aztec room.

The conspicuous artificiality of the Aztec room montage refers us back to a long history of interrelations between architecture, politics, and the Pre-Columbian heritage of Mexico. Politically informed revivals of the Pre-Columbian past were just as prominent in twentieth-century Mexico as they were in Peru, but the instrumentalization of this past and its cultural legacies in Mexico was significantly different. As Marisol de la Cadena has shown, Peruvian intellectuals of such diverse ideological bent as Mariátegui and Víctor Andrés Belaúnde debated the means by which to incorporate Pre-Columbian and indigenous cultures into "national" culture, but did not understand cultural and racial miscegenation, or *mestizaje*, as the primary answer to this question. Conversely, Mexican intellectuals like José Vasconcelos (1882–1959) and Manuel Gamio (1883–1960) embraced various definitions of *mestizaje*, promoting them as key to the construction of an official "Mexican" culture.[30]

Ideas of cultural hybridity received official endorsement during the first two decades after the revolution of the 1910s, informing state-sponsored architecture. As Luis Carranza has demonstrated, Vasconcelos held considerable sway as an architectural patron. As Secretary of Public Education, in 1922 Vasconcelos sponsored the modernization of this institution's host building in Mexico City—a building originally from 1639—refurbishing it with murals glorifying Mexico's racial and cultural mixing by painters Roberto Montenegro and Diego Rivera. During his short but intense tenure at the Secretaría de Educación Pública (SEP) between 1921 and 1924, Vasconcelos touted colonial architecture as a genuine expression of the mixture of European and Pre-Columbian cultures in Mexico, promoted the construction of neocolonial buildings for official institutions, especially schools, and hailed murals as instruments of public indoctrination.[31] The formal and

4.09 "Visit to the Aztec Room," 1964

symbolic appropriation of Pre-Hispanic architecture was also present alongside
neocolonial revivals during the 1920s. This trend was famously exemplified by the
1929 Mexican pavilion at the Seville Ibero-American exposition, where the Yucatan-
based architect Manuel Amábilis produced a thoroughly modern building of
recognizably Mayan features, reviving a tradition of Neo-Pre-Columbian pavilions
that Mexico had presented at several late nineteenth-century expositions.[32]

Revivals of Mexico's Pre-Hispanic heritage went hand in hand with aggressive
support of public education initiatives during the administrations of Lázaro
Cárdenas (1934–1940) and Manuel Avila Camacho (1940–1946). As mentioned
above, Ramírez's career began as a result of state-backed initiatives in education
in 1944, when the architect first became involved with the construction of schools.[33]
Ramírez's practice as head of the CAPFCE, an agency that stemmed directly from
the SEP's early efforts, would also remain tied to debates about the place that
Mexico's colonial and Pre-Columbian pasts should occupy in contemporary culture,
as would his architectural work. Although his first schools did not quite stray from
Functionalist tenets of design, by the early 1950s Ramírez's architecture began to
explicitly revive aspects of Pre-Columbian culture. For instance, the mural at his
1953 National School of Medicine, made out of colored mosaics by artist Francisco
Eppens, monumentalized racial and cultural hybridity. The central element of the
mural is a face that emerges from the fusion of two other faces seen in profile, and
which flank it on either side: the faces of an "Indian" mother, on the left, and of a
"Spanish" father, on the right. The result of the fusion, a face that represents the
modern, *mestizo* nation of Mexico is itself borne out of the fusion of Pre-Columbian
deities, among them the central Mexican feathered serpent, the rain deity Tlaloc,
and the Aztec deity Coatlicue.[34] Built of reinforced concrete, raised on *pilotis* and
sporting a large *brise-soleil* facade, nothing else about the building was explicitly
"Mexican," but its mixture of internationalist aspirations and "national" flavor—a
hybrid condition that defined the entire University City—represented a highly
politicized exploration of *mestizaje* through built form.

Like the readers of Belaúnde's *Pueblo por Pueblo*, visitors to Ramírez's Museum of
Anthropology also engaged in travels through national space, but here the nation
was narrated spatially through a sequence of dramatic environments. Located in
the newly revamped Chapultepec Park, Ramírez's Museum, sponsored by CAPFCE
and the National Institute of Anthropology and History (INAH), comprised twenty-
six rooms devoted evenly in number between didactic displays about Mexico's
Pre-Hispanic history, and displays of contemporary ethnography. The rooms were
organized in relation to a central courtyard, the centerpiece of which was a
monumental umbrella clad in aluminum and supported by a column of carved
bronze that narrated episodes in Mexico's modern and Pre-Columbian history,
made by sculptor José Chávez Morado.[35] An ambitious series of murals by living
Mexican artists dominated several of the Museum's spaces. One of the
introductory rooms included a mural by Jorge González Camarena that, as the
Museum's official publication claimed, "shows the gradual fusion of all peoples
toward a single universal race, or *mestizaje*," and thus presented a sanitized
reinterpretation of the narrative of miscegenation first constructed by Vasconcelos
in his paradigmatic 1925 essay, *La Raza Cósmica*.[36] In his mural for the Museum,
established official painter Rufino Tamayo also focused on *mestizaje*. Tamayo's
Duality featured the epic battle between Pre-Columbian deities Quetzalcoatl and
Tezcatlipoca. Suspended in tension, the image of their confrontation repurposed
a central Mexican myth of creation, presenting it as emblematic of the birth of
modern Mexican national identity.[37]

While a voracious pursuit of international expertise defined the Museum's
planning, *mestizaje* arguably defined its architecture.[38] Ostensibly like institutional

4.10 (left) Manuel Amábilis,
Mexican Pavilion, Ibero-
American Exposition,
Seville, detail of "Mayan"
columns, 1929

4.11 (below) Pedro Ramírez
Vázquez, with Ramón
Torres and Héctor
Velázquez, School of
Medicine, UNAM, Mexico
City, 1953. Mural by
Francisco Eppens

4.12 Pedro Ramírez
Vázquez, with Ramón
Torres and Héctor
Velázquez, School of
Medicine, UNAM, Mexico
City, detail of *brise-soleil*
façade, 1953

4.13 Pedro Ramírez
Vázquez, Rafael Mijares,
and Jorge Campuzano,
National Museum of
Anthropology and History,
Mexico City, 1964

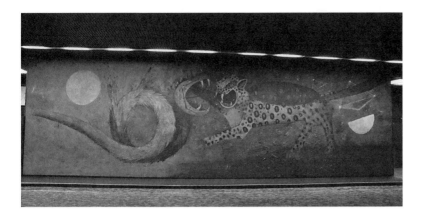

4.14 Rufino Tamayo, *Dualidad*, 1964, installed at the National Museum of Anthropology and History, Mexico City

4.15 Manuel Felguérez, sculpted courtyard facades, National Museum of Anthropology and History, Mexico City

buildings found in many world capitals during the 1960s, Ramírez's Museum incorporated references to Mexican colonial and Pre-Columbian building practices. The use of marble in tandem with *recito*, a rough Mexican volcanic stone used in colonial building, defined many of the Museum's surfaces, including its walls and patio floor. Sculpted metal surfaces designed by artist Manuel Felguérez, which recalled the stone ornamentation of Mayan buildings in the Yucatan peninsula, defined the inner facades of the courtyard. For his part, Ramírez described the patio at the Museum as paying homage to Mayan architecture, specifically to the creation of "quadrangles," public spaces with several access points, in monumental complexes.[39]

The Museum's rooms were designed as independent pavilions, but the hierarchy between them was clear. "Fundamentally," Ramírez wrote, "the layout of the museum can be reduced to a single rectangle running from the outer courtyard to the Mexican room," the entrance to which was marked by a pool that "recall[ed] the lake origins of [Aztec] culture."[40] Iker Larrauri designed the Aztec room's layout. Located at the easternmost section of the Museum's floor plan, visitors to the Museum converged in this chamber from every direction. Spatially, the succession of Pre-Hispanic cultures of Mexico culminated in this bombastic display of Aztec culture, much to the chagrin of critics of the museum since the 1960s.[41] But here too, it was a culturally hybrid space that Ramírez's museum presented. With no columns, the Aztec room was two stories high and included only two partition walls at each side. The walls were only partially load-bearing and spanned only part of the room's depth, giving the impression that the room's tall ceiling hovered over them. Larrauri claims that this effect was intended to recall—and, in his words, "Mexicanize"—a number of international modernist tropes in the construction of display spaces. These spaces created the dynamic effect of abstract, moving planes in space, but here the contrast between such smooth spatial plays and the rugged surfaces of heavy Aztec monoliths was intended to evoke a sense of Mexican cultural specificity.[42] Featured prominently in the center of the room, flanked by a number of Aztec monoliths on pedestals and lit dramatically, the most important work shown in this spectacular space was the Aztec Calendar Stone. Museified in similar fashion as Belaúnde's Stone of Sayhuite in *Pueblo por Pueblo*, here the Calendar Stone became simultaneously enshrined as the essence of Mexico's past glories, and as the crux of Mexico's modern-day imagined identity.

Significantly, Larrauri's Aztec chamber preserved the preeminent position that the Aztec Calendar had occupied in the layout of the Gallery of Monoliths of Mexico City founded in 1877, during the rule of Porfirio Díaz, and hence arguably preserved some of the artifact's political mystique.[43] Yet there was a more recent precedent for this all-important display room, one found in Ramírez's first museum building in Mexico City: the Gallery of the Mexican People's Struggle for Independence, inaugurated in 1960 to commemorate five decades of the 1910 revolution. This snail-shaped structure was built at the side of the hill leading up to Chapultepec Castle, once Emperor Maximilian's residence and the residence of Mexican presidents until 1939, when Lázaro Cárdenas decreed that it become Mexico's National Museum of History. Ramírez's exhibition, scripted by Torres Bodet, targeted an audience of children and included dioramas of heroic episodes in the history of Mexico's transition toward independence from colonial rule.[44]

Although the Gallery mostly focused on Mexico's modern history, its spatial narrative was nothing if not "archaeological" in its language. The Gallery's three levels of displays spiraled down to a cave-like chamber designed by Larrauri, a chamber faced in rough volcanic stone, as deep as the museum's three stories, and animated by a striking confrontation between objects. At one end of the

4.16 Pedro Ramírez Vázquez, Galería de Historia, Chapultepec, Mexico City, 1960

chamber was a copy of the 1917 Mexican constitution—the first republican constitution—attached to a Mexican flag and to the sculpture of a snake's head. At the opposite end of the room was a pillar of sculpted marble, shaped as an eagle slaying a snake and carved in an especially rough fashion to resemble Pre-Columbian stonework by Chávez Morado, who also designed an iron gate for the Museum. The chamber was capped by a circular dome of translucent fiberglass, and was lit so dramatically as to take on a religious aura, rendering visitors silent in the presence of all the ingredients of the Mexican coat of arms. Explicitly inspired by Frank Lloyd Wright's Guggenheim Museum in New York (1949), the Gallery repurposed its spiral form, translating its surfaces and narrative to get across a "nationalist" message. Spectacularized at the Gallery was an encounter of modernist design sensibilities and a Primitivist understanding of identity and history, a combination of elements that had defined Mexican official architecture since the beginning of the twentieth century, and which also defined much of Ramírez's career as state architect in the 1960s.

Conclusion

Mexico's Museum of Anthropology and the Chapultepec Gallery shared their propaganda aims with a set of museums of culture, which included a significant emphasis on Pre-Columbian art, planned for cities along the US-Mexico border, only one of which was eventually built in Ciudad Juárez in 1962.[45] The Museum must also be inscribed within a panorama of significant international exposure of Pre-Columbian art from Mexico, not least in the international pavilions that Ramírez's firm designed for the World's Fairs of 1958, 1962, and 1964–1965, in tandem with the Museum. These experiences paved the way for what may be considered the high point of Ramírez's career in terms of political visibility: his controversial role as the chief organizer for the 1968 Mexico City Olympics, an event that solidified his position as the official architect of Mexico's international image. With an unprecedented focus on cultural display in tandem with the organization of sports events, the discursive framework for the Olympic campaign grew directly out of Ramírez's experience in the design of temporary exhibitions and museums of culture, especially his experience as chief architect of the Museum of Anthropology.

Yet Ramírez's Olympics are not remembered as a moment of triumph, but as a point of rupture between the claims for increased political participation in the Mexican political sphere—especially on the part of a militant student movement in Mexico City—and the reductive elaborations of this identity by an increasingly repressive single-party state. On October 2, 1968, merely ten days before Olympic celebrations started, tensions boiled to a bloody conclusion, as state forces murdered hundreds of protesting students at the Plaza of the Three Cultures in Mexico City. Curiously enough, the Plaza and the Museum of Anthropology shared not only common designers—anthropologist Ricardo de Robina designed both the Museum's general layout, and the Plaza, in 1964—but also told a very similar spatial narrative. The Plaza was made up of a set of carefully manicured Aztec

4.17 Ricardo de Robina, head architect, Plaza de las Tres Culturas, Mexico City, 1964

ruins, a colonial church, and Ramírez's own Ministry of Foreign Affairs building (1966), and was nestled within Mario Pani's Nonoalco-Tlatelolco housing project (1964), the largest ever built in Mexico. Just as the Aztec room operated as the Museum's most significant social space, so did the Aztec ruins of Tlatelolco, a twin city to Tenochtitlán until the sixteenth century, provide the symbolic hearth for Pani's complex, which was nothing if not a monument to Mexico's developmentalist state. Although the massacre at the Plaza cannot be explained merely as a confrontation between a repressive state and the liberal forces that animated the student movement, the confrontation did involve a rejection of a national image projected by the state, an image spatialized by the Plaza itself.[46]

In a striking coincidence, the early days of October 1968 also brought Belaúnde political misfortune. The end of Belaúnde's first administration took place in the midst of a corruption scandal involving the International Petroleum Company, a subsidiary of Standard Oil of New Jersey, a US corporation with a long history of involvement in Latin American politics.[47] The Belaúnde administration's lack of transparency in settling a legal dispute with the Company over the exploration rights for two northern Peruvian oilfields spurred the scandal: after agreeing to pay the corporation for handing the oilfields over to state control, Belaúnde tried to conceal the agreed-upon compensation sum from public view. This sparked widespread outrage and led to the resignation of the greater part of his executive cabinet on October 1, 1968. Belaúnde, whose legitimacy was frail after his first years in office failed to bring the spiritual renaissance or economic boom that his campaign promised, was ousted from office in a coup staged by army general Juan Velasco Alvarado the following night.[48]

Beyond these temporal and biographical coincidences, this chapter has demonstrated that profound historical connections existed between the careers of Belaúnde and Ramírez Vázquez. Providing us with a complex picture of how architecture, politics, and culture interact, the under-examined trajectories of these two architect–politicians compel us to move past the reductive practice of Latin American architectural history as the study of buildings and their heroic patrons and makers. As we have seen, in Peru and Mexico, two "emerging" countries, architects helped enshrine Pre-Columbian artifacts and spaces as key ingredients of evolving and unstable national mythologies, inscribing their efforts within developmentalist campaigns of public education, infrastructure buildup, and economic transformation. A fuller account of these trajectories may help us reformulate the narratives of efflorescence and decline that still define studies of Latin American modernism, while helping us understand Latin American developmentalisms as complex cultural, as well as economic and political, phenomena. This approach may eventually facilitate a more nuanced understanding of the relationships between the spaces, images, and practices of power to which architects in the region often lent a helping hand.

Notes

1   As Jorge Francisco Liernur has demonstrated, some of these architect–politicians were active as members of early twentieth-century vanguards, later rising to prominence in the Cold War as "experts" within the period's newly created institutions. See Jorge Francisco Liernur, "Vanguardistas versus expertos. Reconstrucción europea, expansión norteamericana y emergencia del 'Tercer Mundo,'" *Block* 6 (March 2004): 18–39.

2   The literature on developmentalism is vast. For a broad, and recent account of developmentalism in three key contexts, including Mexico, see John Minns, *Politics of Developmentalism: The Midas States of Mexico, South Korea, and Taiwan* (New York: Palgrave Macmillan, 2006). For a discussion of Belaúnde's first administration, see Pedro Pablo Kuczynski, *Peruvian Democracy Under Economic Stress: An Account of the Belaúnde Administration, 1963–1968* (Princeton: Princeton University Press, 1977). For

a somewhat tendentious, if apt, account of Belaúnde's role in the broader picture of developmentalism in Peru, see Carlos Malpica Silva Santisteban, *Desarrollismo en el Perú: Década de esperanzas y fracasos, 1961–1971* (Lima: Editorial Horizonte, 1975).

3   I borrow this term from Anthony Vidler's essay, "Architecture's Expanded Field," in *Architecture Between Spectacle and Use*, ed. Anthony Vidler (New Haven: Yale University Press, 2008), 143–154.

4   Miguel Cruchaga, interview with the author, Lima, May 20, 2008. Cruchaga is a nephew and early collaborator of Belaúnde's, and also one of the architects, alongside Emilio Soyer and Miguel Rodrigo, of the Ministry of Fisheries Building in Lima (1970–1975), which eventually came to house Peru's National Museum of Anthropology and History.

5   For an analysis of Belaúnde's magazine, see Antonio Zapata Velasco, *El joven Belaúnde: Historia de la revista El Arquitecto Peruano, 1937–1963* (Lima: Librería Editorial Minerva, 1995).

6   Wiley Ludeña, "Fernando Belaúnde Terry y los inicios del Urbanismo Moderno," in *Construyendo el Perú II: Actas del II Congreso de Historia de la Ingeniería y la Arquitectura en el Perú "Fernando Belaúnde Terry"* (Lima: Universidad Nacional de Ingeniería, 2001), 253. The first institution was the Peruvian Planning Office, founded in 1944. This was soon followed by the Peruvian Institute of Urbanism, the first devoted to the specialized teaching of the discipline in Peru, and two other agencies formed in 1946: the National Housing Corporation (CNV), and the National Office for Planning and Urbanism (ONPU).

7   Josep M. Rovira, *José Luis Sert, 1901–1983* (Milan: Electa, 2000), 157.

8   Ramírez discussed his run for the senate in an interview with Enrique de Anda. See "Entrevista con el Arquitecto Pedro Ramírez Vázquez," in *Pedro Ramírez Vázquez: Imagen y obra escogida* (Mexico City: 1988), 15.

9   Roderic Camp, *Mexican Political Biographies, 1935–1993* (Austin: University of Texas Press, 1995), 578.

10 Helen Thomas, "Colonising the Land: *Heimat* and the Constructed Landscapes of Mexico's Ciudad Universitaria (1943–1953)," in *Transculturation: Cities, Spaces, and Architectures in Latin America*, ed. Felipe Hernández et al. (Amsterdam; New York: Rodopi, 2005), 109–123.

11 As Belaúnde's domestic prestige became increasingly compromised, so did his esteem in the eyes of US diplomats and politicians. See Richard J. Walter, *Peru and the United States, 1960–1975: How their Ambassadors Managed Foreign Relations in a Turbulent Era* (University Park: Pennsylvania State University Press, 2010), 38–61.

12 "Peru: The New Conquest," *Time Magazine*, March 12, 1965. See Fernando Belaúnde Terry, *Peru's Own Conquest* (Lima: American Studies Press, 1965), and *La Conquista del Perú por los Peruanos* (Lima: Ediciones Tawantinsuyu, 1959).

13 Belaúnde, *La Conquista*, 97. The project's messianic potential was emphasized in a 1965 publication in which it was renamed as the "Bolivarian" Forest-Edge Highway, evoking nineteenth-century political liberator Simón Bolívar's ideas about Pan-South-American unity. See Fernando Belaúnde Terry, *La Carretera Bolivariana Forest-Edge de la Selva* (Lima: Ministerio de Fomento y Obras Públicas, Dirección de Caminos, 1965).

14 Interview with Miguel Cruchaga.

15 Luis Dorich, "Fernando Belaúnde Terry y la Planificación en el Perú," in *Construyendo el Perú II* (Lima: Universidad Nacional de Ingeniería, 2001), 241.

16 Fernando Belaúnde Terry, *Pueblo por Pueblo* (Lima: Ediciones Tawantinsuyu, 1960). See also by Belaúnde, *El Perú como doctrina* (Lima: Secretaría Nacional de Acción Popular, 1961).

17 See Alexander von Humboldt, *Researches Concerning the Institutions and Monuments of the Ancient Inhabitants of America, trans. Helen Maria Williams* (London: Longman, 1814). Johann Jakob von Tschudi, *Travels in Peru during the Years 1838–1842*, trans. Thomasina Ross (New York: Wiley & Putnam, 1847).

18 The format of the travel book had also been popular among twentieth-century architects, and this could have been another source of inspiration for Belaúnde. See, for instance, Eric Mendelsohn's lavishly illustrated *Amerika; Bilderbuch eines Architekten* (Berlin: R. Mosse, 1926), or Richard Neutra's *Wie Baut Amerika?* (Stuttgart: J. Hoffman, 1927).

19   George E. Squier, *Peru: Incidents of Travel in the Land of the Incas* (New York: Harper & Row, 1877).

20   Ibid., 555–556.

21   Belaúnde, *Pueblo por pueblo*, 105–106. This and all subsequent translations are by the author.

22   Mary Louise Pratt, *Imperial Eyes: Travel Writing and Transculturation*, 2nd ed. (London; New York: Routledge, 2008), 131.

23   Fernando Belaúnde, interview with Enrique Chirinos Soto, published in *Conversaciones con Belaúnde. Testimonio y Confidencias*, ed. Enrique Chirinos Soto (Lima: Editorial Minerva, 1987), 29. Emphasis in the original.

24   Fernanda Beigel, "Mariátegui y las antinomias del Indigenismo," *Utopía y Praxis Latinoamericana* 6, no. 13 (2001): 43.

25   Fernanda Beigel, "El proyecto socialista de José Carlos Mariátegui," in *Itinerarios Socialistas en América Latina*, ed. Estela Fernández Nadal (Córdoba: Alción Editora, 2001), 94–95.

26   Francisco José López Alfonso, "Introducción," in *Indigenismo y propuestas culturales: Belaúnde, Mariátegui y Basadre*, ed. Francisco José López Alfonso (Alicante: Generalitat Valenciana; Instituto de Cultura Juan Gil-Albert, 1995), 30–31.

27   Victor Andrés Belaúnde, *Peruanidad* (Lima: El Mercurio Peruano, 1947).

28   For an archaeology of this strand of Peruvian elite discourse, see Cecilia G. Méndez, "Incas Sí, Indios No: Notes on Peruvian Creole Nationalism and its Contemporary Crisis," *Journal of Latin American Studies* 28, no. 1 (February 1996): 197–225. Also see Mark Thurner, "Peruvian Genealogies of History and Nation," in *After Spanish Rule: Postcolonial Predicaments in the Americas*, ed. Mark Thurner and Andrés Guerrero (Durham; London: Duke University Press, 2003), 141–175.

29   "Peru: The New Conquest."

30   Marisol de la Cadena, "Silent Racism and Intellectual Superiority in Peru," *Bulletin of Latin American Research* 17, no. 2 (May 1998): 150.

31   Luis Carranza, *Architecture as Revolution: Episodes in the History of Modern Mexico* (Austin: University of Texas Press, 2010), 29.

32   Ibid., 102–114.

33   Ramírez was not the only architect to begin his career in the context of post-revolutionary school building. For instance, in 1932, Juan O'Gorman had made his mark in the architectural scene through his design of a set of now-renowned Functionalist schools. See J. Víctor Arias Montes et al., eds., *Juan O'Gorman, Arquitectura Escolar 1932* (Mexico City: Universidad Autónoma Metropolitana and UNAM, 2005).

34   *Guía de Murales de la Ciudad Universitaria* (Mexico City: UNAM, 2008), 66. Although the School of Medicine's mural was among the most ambitious of the University City, collectively the murals harked back to the aggressive promotion of mural painting begun in the 1920s.

35   With architects Rafael Mijares and Jorge Campuzano, Ramírez supervised the work of a large network of museum designers and anthropologists responsible for the display. The Museum's official publication stressed the national magnitude of the operation. "In less than eight months seventy ethnographic expeditions covered the country," gathering materials intended to explain just where the Pre-Hispanic past of Mexico fit within the image of an "emerging" nation. Luis Aveleyra, "Archaeological Research," in *The National Museum of Anthropology*, 33.

36   Ibid., 46. See José Vasconcelos, *The Cosmic Race*, trans. Didier T. Jaén (Baltimore: Johns Hopkins University Press, 1997).

37   Mary K. Coffey, "'I'm Not the Fourth Great One': Tamayo and Mexican Muralism," in *Tamayo, A Modern Icon Reinterpreted*, ed. Diana Dupont (Santa Barbara: Santa Barbara Museum of Art, 2007), 260.

38   In February 1961, for instance, Luis Aveleyra, an anthropologist who curated the Museum display, wrote to sound engineer Eugene Miller about acquiring an audio guide system very recently used at the American Museum of Natural History in New York. Letter, Luis Aveleyra de Anda, to Eugene Miller, February 15, 1961. Archivo Histórico, Museo Nacional de Antropología, Mexico City, Vol. 185, Folio 80. Museum designer Alfonso Soto Soria recalled that the Smithsonian Museum in Washington, DC, which he saw as part of an official visit, provided examples of non-reflective glass

casing and lighting that were later used. Alfonso Soto Soria, interview with the author, Texcoco, August 28, 2009.

39 Ramírez Vázquez, "The Architecture of the Museum," in *The National Museum of Anthropology*, 20.

40 Ibid.

41 See the famous critique by Octavio Paz, *The Other Mexico: Critique of the Pyramid*, trans. Lysander Kemp (New York: Grove Press, 1972), 108–110.

42 Larrauri, interview with the author, Mexico City, August 23, 2009. Larrauri cites as sources of inspiration the exhibition designs of Mies van der Rohe, especially Mies's 1929 Barcelona Pavilion.

43 See Khristaan D. Villela and Mary Miller, eds., *The Aztec Calendar Stone* (Los Angeles: Getty Research Institute, 2010).

44 Pedro Ramírez Vázquez, *Ramírez Vázquez en la Arquitectura* (Mexico City: Universidad Nacional Autónoma de México; Editorial Diana, 1989), 80.

45 Bruce Munro,"PRONAF," in *Mexico and the United States*, ed. Lee Stacy (Tarrytown: Marshall Cavendish Corporation, 2003), 672–673.

46 For a fuller account of this relationship, see my "Beyond Tlatelolco: Design, Media and Politics at Mexico '68," *Grey Room* 40 (Summer 2010): 100–126.

47 See Charles T. Goodsell, *American Corporations and Peruvian Politics* (Cambridge: Harvard University Press, 1974); Adalberto Pinelo, *The Multinational Corporation as a Force in Latin American Politics: A Case Study of the International Petroleum Company* (New York: Praeger, 1973).

48 "Fue derrocado presidente Belaúnde esta mañana. Una Junta Revolucionaria asume el poder," *El Comercio*, October 3, 1968. Velasco ruled as Peru's de facto president until 1975, when army commander Francisco Morales Bermúdez deposed him. In 1979, Morales called for new elections; Belaúnde won the elections, held in 1980, to serve a second term in office.

5

CARACAS'S CULTURAL (BE)LONGINGS

Retracing the Troubled Trajectories of
the *Superbloque* Experiment[1]

Viviana d'Auria

As a significant component of the rich scenario encompassed by modern Latin American architecture and urbanism, the Caracas *superbloques* (1951–1958) did not escape the contradictory processes of appreciation and rejection that they provoked, amid a complex intertwining of political, economic, and cultural factors both internal and external to the region.[2] In general terms, the residential developments designed by the Architectural Office of the Banco Obrero (TABO)[3] reflected widespread efforts to devise modern solutions to house the urban dwellers of a rapidly changing world. In Venezuela, as part of the oil-lubricated impetus toward modernization, these intensive high-rise neighborhoods both supported the production of a particular version of national identity and secured Caracas a position among the top cities of modern architecture worldwide. In the *superbloques*, modernity and modernization collided: high-rise technologies, cultural and socio-political emancipation met a new urban scale liberated from all gridded, colonial constraints.

The government-funded high-rise interventions can be examined as some of Caracas's most significant urban topographies, acting as a repository for ambivalent cultural yearnings and disputes. The TABO residential experiments offered a rich framework for postwar practitioners to debate the longevity of cultural influences from across the Atlantic and more recent exchanges with the United States. Developments such as Cerro Grande, El Paraíso, and 23 de Enero have acted as urban constructs attesting to the ever-increasing interpenetration of local and global features in their conception, contestation, and appropriation, making suspect "European" and "(Latin) American" categorizations of "imported" modernity. The intermingling of references involved in Caracas's construction points to a transcultural city-making engendered by the coexistence of "entangled" modernities and by no means following a one-way process.[4]

The maelstrom of critiques the *superbloques* have triggered reveals the powerful projections superimposed upon Caracas during a period of substantial geopolitical realignment. From Henry-Russell Hitchcock's celebratory descriptions to the CINVA (Inter-American Housing Center)[5] evaluation leading to the suspension of all *superbloque* construction in Venezuela, reviews of TABO's "crash program" are located at the nexus of Cold War transnational concerns about the urban home and Latin American modernization and state-building. Contemporary reassessments argue instead for a reappraisal. Their claims are rooted in the *superbloque* neighborhoods' supposed potential to support insurgent social identities, often generating a disquieting aftertaste of misplaced agency. Indeed, while James Holston has defined "insurgent citizenship" as a destabilizing counter-politics that refuses the simplistic association of citizens' agency with heroic resistance, the epic nature of *superbloque* everyday life has instead been reified by several architects recently engaged with the housing complexes.[6] In so doing, they risk misrepresenting the inhabitants' activism as an exceptional condition requiring celebration, rather than as an essential blend of institutional and extra-institutional mobilization inherent in the democratic life of the *superbloques* and their auto-constructed surroundings.[7]

5.01 (above) *Superbloques* as the new urban scale. Carlos Raúl Villanueva, Guido Bermúdez, and Carlos Brando for TABO, Cerro Piloto, Caracas, 1954

5.02 (right) *Superbloques* as built: Comunidad 2 de Diciembre (today 23 de Enero). TABO, Caracas, 1956

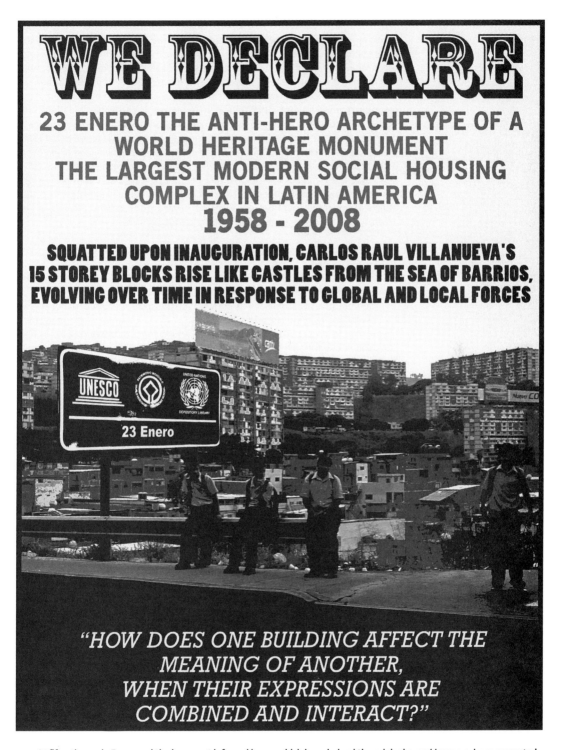

5.03 At fifty, the 23 de Enero and the insurgent informal layers which have helped the original *superbloque* work are presented as a counter-model to UNESCO heritage policy in Caracas. Urban-Think Tank, 11th International Architecture Biennale, Venice, 2008

This inquiry uses the variegated *superbloque* evaluations as an orientation device within a multifaceted and ambiguous postwar panorama, investigating the (dis)location of Caracas on—and off—the maps of the modern urbanizing world, recounting the tale of a contested city hosting several overlapping topographies. It examines how Caracas was projected into the territories of modern urbanism and architecture, plunged back into the realm of (under)development, and recently claimed as a product "made in Venezuela."[8] The analysis underscores both exogenous and endogenous yearnings accompanying the city's postwar development, attesting to the multidirectional and iterative evolution of Caracas's incomplete modern project. Moreover, the *superbloques'* subsequent takeovers point to the complexes' haunting presence within the urban lifeworld they have been part and parcel of for the past half-century. The fact that such projections are re-proposed today under radically different conditions suggests either that binary oppositions could be reinstalled, or that a more refined reading of the environment could be achieved.

Architecture and/or Revolution?

Le Corbusier and CIAM claimed . . . that revolution could be created through architecture. It has been proven that this is not true, that it is not possible, at least not in a situation such as ours. So it [the 23 de Enero *superbloque*] is an interesting experience and I think that in thirty or forty more years we'll be able to draw our conclusions. I hope those conclusions can lead to a new perspective with a deeper understanding of how to really transform a society and how to develop an architecture that is really appropriate and adapted to that new situation. But that's how we see it now. We didn't understand that back then.

(Juan Pedro Posani, 2005)[9]

The Caracas *superbloques* were rooted in a dictator's dream. Eighty-five high-rise buildings constructed at breakneck speed seemed sufficient to conceal, once and for all, the *ranchos* perched on the steep hillsides embracing the narrow valley.[10] In most instances, bulldozer law accelerated the resettlement process, compressing it into a few hours. Petroleum provided the resources for the pace and scale of intervention, which would outstrip other contemporary developments in Latin America, arousing extensive admiration from abroad. The *superbloques* however, were also guided by the vision of a small team of avant-gardist practitioners and students—Carlos Celis Cepero, José Manuel Mijares,

5.04 Discussing housing layouts at the TABO: Carlos Brando, Carlos Raúl Villanueva, Guido Bermúdez, and Carlos Celis Cepero, Caracas, 1951

5.05 The TABO architects at work: Carlos Brando, Victor Mantilla, Carlos Raúl Villanueva and Guido Bermúdez, Caracas, 1955

Carlos Brando, Victor Mantilla, and Guido Bermúdez—that Carlos Raúl Villanueva assembled around him,[11] with Le Corbusier's Atelier des Bâtisseurs as their operational model.[12] The group aimed to "renovate the concept of habitat," particularly that of social housing.[13] Housing was viewed as the key component for restructuring society, particularly the in-migrating communities of the rapidly growing city, and catering for their socio-economic development was the architect's fundamental task. Since public housing was also the place where individual and collective scales met, the question of mass democratization and emancipation meant community service provision and typological variety were prime ingredients for practitioners engaged in the rearticulation of urbanizing society by means of neighborhood and dwelling design. The team's official role as part of the newly founded TABO, created to address the country's pressing shelter problems more vigorously, was to develop housing prototypes supporting the National Housing Plan.[14] Iteration between typological research and project assessment was intended to knit the office's working phases closely together. At its inception a wide range of solutions, including cooperative housing and single-family dwellings, were studied to tackle the general problem of building new houses for Venezuelan families, while in Caracas the reliance on high-rise solutions was considered immediately.[15] Following a 1953 survey on informal settlements, and as housing policy swerved and radicalized into the Presidential Program for the Eradication of Insalubrious Housing in Venezuela, TABO would be compelled to focus on slum clearance and produce projects at a much faster pace.[16]

The mismatch between the brutal ambitions of General Pérez Jiménez[17] and modernist experimentalism emerged as the *superbloque* typology was qualitatively reduced while its application throughout the city increased.[18] Architecture students' thesis work fed directly into TABO's quest to conceive a replicable housing block avoiding monotony and schematic layouts. During the experimental years, multi-cellular prototypes were designed and, having proved their economic

5.06 *Superbloques* and the
question of social housing:
working sketch of
Comunidad 2 de Diciembre
(today 23 de Enero). Carlos
Celis Cepero, Urbanización
23 de Enero, Caracas, 1952

5.07 The preamble to slum
eradication through high-
rise intensification:
superbloques as the solution
to squatting. TABO,
Caracas, 1954

5.08 *Superbloque* studies:
the proximity between
model and reality. Carlos
Raúl Villanueva and Carlos
Celis Cepero for TABO,
Quinta Crespo (unrealized),
Caracas, 1951

feasibility, saw massive implementation. When compared to the team's initial *superbloque* projects exhibited in Caracas for World Urbanism Day in 1951, substantial differences in design can be spotted: balconies and loggias have been eliminated, indoor staircases have replaced external ones, variation in apartment typology has dwindled, and uniformity is counter-balanced though chromatic treatment. Last but not least, only forty-two days were necessary to construct a block. Subsequently, the frenetic and intensive implementation of a type stripped of any attributes ensuring some form of spatial complexity continued relentlessly. Condensed into a cement skeleton concealing its bareness by means of colored elevations, the blocks rose on the one hand as the authoritative gesture of a despot, and on the other as the presumptuous emblem of a socio-spatial makeover heralded by modernist urbanists.

Between 1954 and 1958, 80 percent of the *superbloque* apartments were allocated to those evicted from their *ranchos*. In January 1958, however, after four years of intensive construction, housing production and its revolutionary promise of social betterment was interrupted by political insurrection. As Pérez Jiménez fled from the city he had irreversibly shaped, demonstrators occupied 3,880 units as part of the general upsurge which came to be known as the 23 de Enero. This also became the new name of the largest *superbloque* complex of all, commemorating a decade of struggles against a dictatorship. Turned into an arena for contestation, some of the *superbloques* were thus charged with the responsibility for remembrance.

The Unidad de Habitación Cerro Grande saw the light shortly after its homonymous antecedent in Marseilles was completed. The National Housing Plan's exhibition catalogue accompanying the launch of its 1951–1955 phase paired TABO's guiding tenets with aphorisms from Le Corbusier and Auguste Perret. It was claimed that adherence to modern architectural principles would guarantee a "joy of living," the maximal expression of existence.[19] Multi-story blocks and low-rise walk-up apartments would be combined with community buildings to compose cohesive neighborhood units, their groupings defining the new urban scale of the intervention. The typological mix reflected the CIAM debate on the use of Unité-type units or "transitional" carpet-housing in the diffusion of modernist architecture to various corners of the globe.[20]

The neighborhood unit's ideal dimensions, the buildings' setting within the land-scape, the separation of pedestrian from vehicular flows, the central position of

5.09 Re-founding Caracas through "crash" construction. Carlos Raúl Villanueva, Guido Bermúdez, and Carlos Brando for TABO, Cerro Piloto, Caracas, 1955

5.10 *Superbloque* experiments between Le Corbusier and Auguste Perret. Carlos Raúl Villanueva and Carlos Celis Cepero for TABO, Quinta Crespo (unrealized); Guido Bermúdez for TABO, Cerro Grande. Banco Obrero, *Venezuela: exposición 1951–1955. Plan Nacional de la Vivienda*, Caracas, 1951

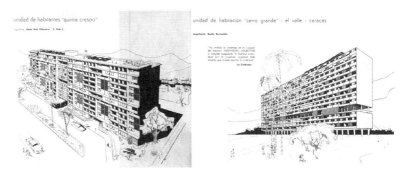

community facilities were the recurrent principles governing the layouts. Maximal liberation of ground space for promenades and landscaping paralleled the ideological liberation from land speculation. Differentiation in block orientation avoided monotony by providing the widest possible range of views. Additionally, grouping rules enhanced interaction between indoor and outdoor activities. Housing units were interspersed according to height differences, and responded to topographical irregularities. The dramatic manipulation of Caracas's hillsides this involved was more than frontier expansion; it came close to re-foundation.

The *superbloques*' lack of finish, crude appearance, and simplified layout did not go unnoticed.[21] Nonetheless, Caracas was positioned loudly and clearly among the cities embracing modern architecture. In his 1955 compilation of recent Latin American architecture, Hitchcock referred to Cerro Piloto as "capable of realizing one of the recurrent dreams of twentieth century urbanism."[22] In other Western publications, the TABO projects were depicted as "cosmopolitan architecture," constructed in a dream-like environment where a well-resourced, highly operational welfare system existed—a condition the European milieu was yet to build for itself.[23]

The renowned *L'Architecture d'aujourd'hui* assembled an atlas of cities championing modern architecture and planning,[24] depicting its diffusion to the rapidly developing and highly receptive regions of the south. Constructing the canon of a modern, International Style rooted in Corbusian thinking was indeed high on the agenda of architectural magazines endorsing the project of modernity. With its status as the "buildingest city in Latin America,"[25] Caracas and its high-rise

5.11 (above) The *superbloques* in progress: freedom from topography, from the grid, and from speculation. Carlos Raúl Villanueva, Guido Bermúdez, and Carlos Brando for TABO, Cerro Piloto, Caracas, 1955

5.12 (left) Caracas among the world capital cities of modern architecture. André Bloc, "L'Urbanisme des Capitales," *L'Architecture d'aujourd'hui* 88, 1960

housing were consistently included in such overviews. In an analogous fervor to prove the international vocation of modernist edifices, theme issues devoted to collective housing showed the *superbloques* side by side with experiments in France, Belgium, the Netherlands, Switzerland, Italy, Sweden, Uruguay, Brazil, Morocco, the United Kingdom, and the United States.[26]

It only took a decade for the bitterness of reality to cut through the sheen of celebratory commentaries. Uncanny panoramas of a valley pressured by population tensions would soon be labeled as typically "Third World." The high-rise panacea had fallen short of its principal pledge. Even though by the end of the decade the *superbloques* housed 13 percent of the capital's population, at least a quarter were still *rancho* residents.[27] The sun had begun to set on the CIAM-inspired project for Caracas. The demise of the *superbloque* was sealed by the publication of the 1959 CINVA-BO evaluation report, which set the trend for the systematic, critical investigation of Venezuela's housing projects.[28] Headed by Eric Carlson, CINVA's interdisciplinary, international team emphasized financial and management considerations. Policy effectiveness rather than architectural quality was on the agenda. As the report itself clearly stated, it had "a marked emphasis on social aspects and social problems . . . caused by the fact that in any basic analysis of economic and technical factors, social questions always arise which usually have never been properly considered."[29] Evidently, the state of economic and social disaster the high-rise housing blocks were in could not escape the assessors' attention: their final recommendations proposed the suspension of all *superbloque* construction until housing policies were defined in relation to the country's economic and social development.[30]

Architectural and technical qualities were unable to reorient attention away from socio-economic difficulties. The *superbloques* were described as "spectacular, brightly painted in multi-colored hues, and arrayed in majestic groupings on hillside sites, with less than 20 percent land coverage."[31] In contrast to the later generic criticisms of high-rise solutions, the concentration of multi-story buildings on the same site did not raise questions concerning the developments' full integration within the urban fabric. Neither was vertical housing seen as disruptive in a city where apartment living was dominant due to topography. Regarded as extremely problematic were elements of a completely different nature: the political extremism of some occupants, the overall social heterogeneity of residents coming mainly from rural areas, their inability to pay rent, and, most significantly, the fact that in-migration to the capital had not been halted. On the basis of this particular ineffectiveness, and the steady growth of *ranchos*, the assessors advised an alternative path.

The report, delivered in April 1959, combined revelatory socio-anthropological portraits of urban in-migration with pleas for an institutionalization of housing policy.[32] Ultimately, the general endorsement of self-help housing and community development were on the agenda.[33] It also reflected the gradual shift from confidence in the transformative capacity of "cosmopolitan" modernism to the fear of social disruption and upheaval engendered by uncontrolled urbanization which, in cities like Caracas, was relentlessly on the increase. The document postulated that a transition was required from traditional to urban lifestyles and that this could be achieved through self-help and community development projects. As the report concluded:

> housing institutions in Latin America and other developing areas must face the fact that for many years to come they will be primarily concerned not only with very low-income families, but with families unaccustomed to urban standards and living conditions, families who have migrated from rural areas

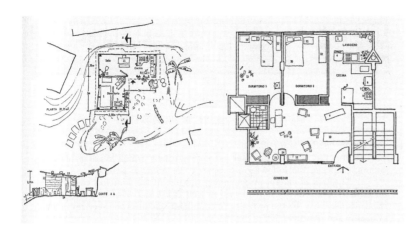

5.13 Plan and section of a typical *rancho* set against a standard two-room apartment in the Simon Rodriguez *superbloque* development. Eric Carlson, for CINVA/Banco Obrero, 1959

unequipped culturally, socially, and emotionally for the "shocks" or for the disciplines imposed by urban life.[34]

The Caracas high-rises became instrumental in consolidating these arguments in favor of self-help (then being incorporated into the institutional wisdom of international aid agencies), and in supporting the application of social sciences to the administration of housing and urbanization. In this respect, the report reflected the distance between CIAM's emphasis on both formal and social components and the more conventional consideration of design as one of the "technical aspects" by CINVA, the region's preeminent center for the dissemination of self-help housing.[35] As Carlson's concerned descriptions of social heterogeneity and political unrest revealed, the *superbloques* failed to provide insurance against the residents' economic insecurity and political volatility. They were also far from a self-help model where shortcuts to ownership (via savings and mortgages), incremental growth, and inhabitants' involvement were the core of housing programs.[36]

The various modernities at play within the making of Caracas thus offered conflicting interpretations of the *superbloques*. That section of the elite craving the immersion of Venezuela into French cultural patterns interpreted the Unidad de Habitación and TABO's avant-gardist approach as the materialization of their long-awaited reconnection to Paris—however untimely this sentiment was, given the increasing "inter-American" influence.[37] On the other hand, the emerging alliances between Venezuelan architects and their US colleagues significantly reoriented these ambitions, through efforts to mold housing policies via substantial economic and technical assistance. Particularly for publicly financed low-income housing, the strategies endorsed by CINVA's report inaugurated an approach that was consolidated via US foreign aid and shaped by Cold War imaginations. Expanding access to private home-ownership, guaranteeing "respectable" living standards, and relying on "core" or "embryo" typologies to facilitate "people-driven" interventions became tools for introducing middle-class habits to less-favored groups. Similarly to the core houses developed as part of US and UN technical assistance programs in the early 1950s, embryo houses (single room, kitchen, and bathroom), connected to broader notions of self-help housing. Both were based on the rearticulation of dwelling units according to homeowners' requirements over time, obtained by unleashing individual and collective energies. Ultimately, mass consumption and moderate political behavior were key objectives.[38]

Successive criticisms by key US "housers," in large part building on Carlson's report, also criticized design-related issues and articulated anxieties vis-à-vis poor management, and the socio-economic problems engendered by the high-rise complexes. Charles Abrams, for example, underscored the divergence—in terms of scale, quantity, and cost—between the publicly financed dwellings and prevailing processes of urbanization, expressing his disapproval that the *superbloques*, as "monumental tablets," proclaimed their superiority to the "make-shift *ranchos*" nearby.[39] Monumentality—set in opposition to socially oriented public service—became the focus of criticism from Francis Violich, who had originally greeted the *superbloques* as one of Latin America's more promising approaches, and had also actively developed housing strategies as a consultant to the National Commission on Urbanism.[40] Similarly, after CINVA-inspired remediation measures were taken, John Turner's positive reception of the *superbloques* was later obscured by a harsher verdict on their costly maintenance and the unbridled densification of their surroundings.[41]

5.14 "Monumental tablets" versus "make-shift *ranchos*." Charles Abrams, *Man's Struggle for Shelter in an Urbanizing World*, 1964

5.15 The "monstrous multi-storys" amid unstoppable urbanization. John F. C. Turner, *Housing by People*, 1976

The CINVA evaluation became "a classic of its kind" within development practice.[42] Numerous studies on Latin American housing and urbanization and on the more general planning problems of "developing" countries would refer to the *superbloque* case, more often than not in anecdotal terms. The original CINVA critique would, on the other hand, seldom be questioned. Quoted mostly in its sections referring to apartment overcrowding, insufficient community facilities, rent delinquency, costly social programs, and overall mismanagement, Caracas's profile as a Third World metropolis was consolidated within the networks that were building a transnational language of developmentalism.[43] In relation to Venezuela itself, the report marked the high-rises' delegitimization on several fronts. Besides their inadequacy as model housing for resettled squatters, an understanding of the *superbloques* as symbols of oppressive power led to the condemnation of avant-garde architecture's compliance with dictatorial rule:[44] the *superbloques* were "symbolic works of the previous dictatorship," and many of their socio-economic and administrative difficulties were "political in character."[45] This critique was all the more ironic given TABO's original underlying ambition to reach beyond cultural renewal and facilitate the democratization of public life.[46]

As two sides of the same coin, the *superbloques* and the *ranchos* continue to generate reflections that underline both their interdependence and their contrasts, particularly since the densification of surrounding open spaces unremittingly persists.[47] On sites such as the 23 de Enero, self-built growth around the *superbloques* has steadily increased until these have become literally embedded within the *ranchos*. By looking closely at such interaction, artists Sabine Bitter and Helmut Weber have drawn attention to contemporary urban restructurings driven by the tension between formal and informal processes. High-rise buildings and *ranchos* are in close contact, their relationship built on permeability. Self-organized community groups and radical activism span both large-scale structures and informal inlay, blurring the contraposition between large-scale modernist blocks and the *barrio*, to the point that the resulting construct is presently known under one designation: "el 23."

Referring to the remarkable processes of urban appropriation and (re)development through which inhabitants previously excluded from political decision-making are now taking the lead, Bitter and Weber point to a middle scale where grassroots initiatives meet top-down governmental reformism. Interventions include the construction of public facilities, formalization of tenure, educational programs, and urban agriculture.[48] The resulting "transformative urbanism" is weaving new sites back into Caracas's urban fabric.[49] The locus where the blocks' super-structure and the fine-grained spontaneous settlements intermingle provides the driving spatial metaphor for the artists' work. Informality stands for resistance, and enables the complete merger of high-rise modernist megastructures and dynamic *rancho* development through the alliance of social categories historically rejected by urban transformation. In such a context, the *superbloque* has become a "core" operating at the urban scale, providing nourishment for the surrounding self-organized urban fabric which benefits from its topographical projects and infrastructural networks. Structures imagined as disjointed became instead a cohesive environment sharing the destiny of low visibility within formal city-making. The *superbloques*' afterlife thus underscores the considerable collective effort expended to provide what housing policies did not, due to a lack of insight or a lack of history in doing so.

The outcome of the interplay between grandiose state developments and the ever-evolving *ranchos* is reflected in community-based initiatives addressing what are considered the marginalized sectors of society. Among these is the Museo de los Barrios 23! where the liveliness of this interaction is again heralded as the

5.16 *Superbloques* and their hereafter: vertical *barrios* or living megastructures? Sabine Bitter and Helmut Weber, 2003

basis for architectural conception. Designing the museum has been considered as an exercise in self-representation elaborated by a site which has scarcely had the opportunity to express its location within the metropolis. As the first official structures within which this dialogue has occurred, the *superbloques* hold a particular place in the museum's presentation.[50] As with Bitter and Weber, the synthesis of informal dwellings and formal high-rise into a radical urban structure is taken as an example to reassert the presence of areas within the city that are typically ignored. Ultimately, the roof of Block 26b was selected by the community as the museum's preferred location because of its consolidated reputation as a refuge against persecution by the police.[51]

Revealingly, the interplay between *superbloque* and informal settlements has led to nick-naming the former as "vertical *barrios*."[52] This amalgamation has fostered contemporary architectural speculations that move on from previously imagined interventions addressing the high-rise projects.[53] Envisaging the scenario of further *rancho* densification, the interdisciplinary design practice Urban-Think Tank has proposed sturdier alternatives to the hazardous accumulation of spontaneous settlements along Caracas's steep hillsides. "Vertical *barrio*" is also the first appellation of their multi-level, raw concrete frame with accessible connections and self-built units inspired by both the generative growth of slum-building and the robustness of the high-rises.[54] Amended by the epistemological contribution of informality, more porous than the *superbloques* and more resistant than the *barrios*, the structure is meant to intertwine the tale of both realms, though underpinning the constant game of contraposition and fusion between them.

5.17 (below left) Striving for self-representation in "el 23": Museo de los Barrios 23! Urban-Think Tank/ Kulturstiftung des Bundes, Caracas Case Project, 2003

5.18 (below right) The "growing house" as the best of both? Urban-Think Tank, Caracas, 2003

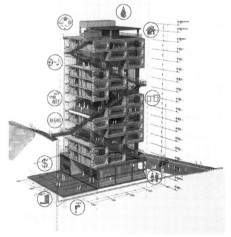

Consequently, the "vertical barrio" or "growing house" still sets questions about the perpetuation of the formal/informal dichotomy which can only with difficulty be solved by simplistically merging the two terms of a binary opposition.

Recent narratives involving the high-rise developments, therefore, not only recognize the housing's relevance for re-articulating architects' terms of reference, but also generate critical questions about the persistence of a dualistic understanding of (in)formality and an ensuing "urban orientalism" concerning the self-built city.[55] Architects' structural engagement with the latter was long overdue. However, unfolding insurgency is not to be approximated with practicing within the auto-constructed environment or being inspired by its self-organization. Moreover, the ways in which the housing born as the remedy to unauthorized city-making has recently been celebrated as part of an exciting architectural do-it-yourself approach is fairly paradoxical: rooted as they were in the formal-informal dichotomy, the *superbloques* have presently come to embody the triumphant "everyday" of "informal" environments. The binary opposition finds itself inverted but also perpetuated, and the challenge for more refined interpretations is left unreciprocated.[56]

Epilogue: The Vicissitudes of Re-Mapping Oneself

> Un superbloque es lo mejor para poder vivir, no vengas a decir que hay algo
>   superior,
> Porque soy pana de Julián, soy llave de Martín, así como verán soy chévere de
>   aquí.
> Por eso busca un novio de otro lote, yo no te cambio por mi superbloque.
>   (Simón Díaz)[57]

Heroic, cosmopolitan modernism, Third World developmentalism, and contemporary activist architecture have all rooted their claims in the high-rise complexes of Venezuela's capital. How are these three overlapping *superbloque*-based topographies informative of where Caracas has been (dis)located to and how it is repositioning itself through the joint venture of state-driven urban transformations and self-regulated city-making? Can binary oppositions be circumvented, allowing the dismantled silhouettes of Caracas's "modernist ruins" to break into a plurality of histories set against the changing and unconfined city?[58] Where do arguments concerning the "spectacular failure" of the Caracas *superbloques* stand with respect to the conception, contestation, and appropriation of these complex multi-story buildings through time?[59]

After the oppressive legacy of *perezjimenismo* and modernist utopianism, most of Caracas's high-rise complexes have experienced not only further episodes of state coercion, but also the vicissitudes of insurgent urbanisms and the stigmatization entailed by political resistance. If this uneasy layering can in part be attributed to the *extranjerización* resulting from the elevated degree of foreign influence on 1950s modern architecture,[60] it also relates to the high-profile seizures impressed on the *superbloques* by those wishing to intensify and exacerbate a dualist interpretation. However disruptive these takeovers may have been, the *superbloques*' complex environment is now involved in a substantial effort at recoding. Laden with half a century of inhabitation, it may start articulating—through rehabilitation and legitimized self-organization—the fragmented polysemy of multiple narratives.

A similar recoding extended to the vicissitudes of Latin American modernity in the context of Caracas can also be envisaged. By recognizing foreignness, alienation, and dislocation as intrinsic to the modern project, tensions inherent to the unevenness of cultural relations between different locales involved in modernity and modernization's expression reverberate back to each party. TABO's "transcendental" objectives of extending modernity to society's large majority entailed drastic typological reduction and neglected the complexity of housing allocation.[61] The immediacy of CINVA's critique, largely intermingled with a broader attack on Pérez Jiménez's dictatorship, shattered the voluntarism of solving mass housing problems through an inclusive, instant state welfare.

Implemented on the back of the oil boom, intensive high-rise housing for the masses came at a time when paradigms were shifting and both US and UN technical assistance in Venezuela was growing. In Caracas, as elsewhere, the unresolved issue of the dynamics and agency of ownership preoccupied social scientists; they gave little support to large-scale architectural projects, advocating instead for other solutions altogether. State-funded housing soon gave way to the promotion of self-constructed, incremental growth, with gradual integration into Venezuela's urban environment.[62] The self-help upgrading programs implemented during the 1960s and 1970s generated the socio-environmental conditions facilitating the massive densification of Caracas's hillsides, without any efforts to change the situation from which housing deprivation originated.[63] Nonetheless, the BO's "outstanding results" and "spectacular progress" in the *superbloques*' administration demonstrated the potential for solidarity between intensive public housing provision and community development projects strongly cherished by CINVA.[64]

However distinctive in their settings and goals, the three topographies discussed emphasize the crucial role of the high-rise projects for advancing contrasting agendas at both local and international levels. From harbingers of modernity to symbols of oppression and icons of resistance, projections of the *superbloques* have yet to strike a balance between social reformism's utopian vigor, radical community-driven initiatives, and romanticized contestation. More than self-build and self-help, and beyond the years of self-management prompted by survival strategies, it is today's struggle for self-representation which might allow the *superbloques*' imago—in its full maturity—to reposition its presence within Caracas and beyond.

Notes

1   Gratitude goes to Paulina Villanueva and the staff of the Carlos Raúl Villanueva Archives, Caracas. I thank Carlos Celis Cepero for his vivid reconstructions of the TABO, as well as Mateo Pintó for sharing his work on the *superbloques*. Generous input from Juan José Perez Rancél and his CIPE colleagues at the Facultad de Arquitectura y Urbanismo (FAU), Universidad Central de Venezuela (UCV) have also been fundamental. Bruno De Meulder's comments at the Katholieke Universiteit Leuven have also been vitally inspiring.
2   See Valerie Fraser's questions concerning this fundamental shift from encomium to opprobrium in *Building the New World: Studies in the Modern Architecture of Latin America 1930–1960* (London; New York: Verso, 2000).
3   The Banco Obrero (BO) was created in 1928 as a financing agency to provide loans for workers' housing. In 1941, it was charged with addressing Venezuela's shelter problem, and the TABO was established in 1951. The TABO ended in 1958, with the fall of Pérez Jiménez and Villanueva's abandonment of the BO. See Banco Obrero, *Sesenta años de experiencias en desarrollo urbanístico de bajo costo en Venezuela* (Caracas: INAVI, 1988); Carlos Celis Cepero et al., "Conversación sobre el TABO," *Revista Punto*, no. 66–67 (1996): 73–80.
4   Though this investigation agrees with the prominence of French influence on Venezuelan urbanism in the 1940s and 1950s, it supports Lu's plea for more complex frameworks for analyzing modernism. See Duanfang Lu, "Entangled Modernities in

Architecture," in *The Sage Handbook of Architectural Theory*, ed. Greig Chrysler et al. (forthcoming); Duanfang Lu, *Third World Modernism: Architecture, Development and Identity* (London: Routledge, 2010). On the question of "transculturation" see Felipe Hernandéz et al., eds., *Transculturation: Cities, Spaces and Architectures in Latin America* (Amsterdam; New York: Rodopi, 2005).

5   CINVA was established upon the initiative of the Organization of American States in 1951 in Bogotá, Colombia as a regional center for training architects, engineers, and social workers in low-cost housing. To date the CINVA archives have not been traced.

6   See James Holston, *Insurgent Citizenship: Disjunctions of Democracy and Modernity in Brazil* (Princeton: Princeton University Press, 2008).

7   See Alejandro Velasco, "'A Weapon as Powerful as the Vote': Urban Protest and Electoral Politics in Venezuela, 1978–1983," *Hispanic American Historical Review* 90, no. 4 (2010): 661–695.

8   Sabine Bitter and Helmut Weber, *Caracas, Hecho en Venezuela* (Frankfurt: Revolver, 2005).

9   Juan Pedro Posani, quoted in Bitter and Weber, *Caracas*, 21. As Villanueva's long-time collaborator since 1949, Posani co-founded the journal, *A: Hombre y Expresión*. He also actively participated in establishing architectural history and theory courses at the FAU, UCV.

10 *Ranchos* are the name given to informal dwellings in Caracas. The term originally denoted a farmer's house, but was transplanted to the city with rural–urban migration.

11 Europe-born and trained, Villanueva studied architecture at the École des Beaux-Arts, Paris. His relocation to Caracas in 1928 was his first prolonged experience of Venezuela. In the 1940s, he began his twenty-year experimentation in public housing within the BO. Colombian architect Carlos Celis Cepero joined the TABO after collaborating with Le Corbusier and Josep Lluís Sert on the Master Plans for Bogotá, Medellin, and Cali. FAU students and artists such as Mateo Manaure and Carlos Bogen also contributed. The TABO's avant-gardist spirit paralleled that of anti-establishment artistic groups like the Disidentes.

12 Celis Cepero et al., "Conversación," 74.

13 Carlos Celis Cepero, interview with the author, March 2009.

14 The National Housing Plan (1951–1955) envisaged the construction of 12,185 residential units. See Carlos Raúl Villanueva and Carlos Celis Cepero, *La Vivienda Popular en Venezuela: 1928–1952* (Caracas: Banco Obrero, 1952), unpaginated.

15 Banco Obrero, *Venezuela: exposición 1951–1955. Plan Nacional de la Vivienda* (Caracas: Banco Obrero, 1951). The first *superbloque* designs were isolated middle-class developments within the city. Quinta Crespo (unrealized), Cerro Grande (1951–1954), and El Paraíso (1954–1955) maintain most of the Corbusian model's salient features, including its appellation, "unidad de habitación." Diego de Losada (1953–1954) and Cerro Piloto (1954), designed for low-income dwellers and without integrated facilities, served as prototypes for the most intensive interventions, 2 de Diciembre (later renamed 23 de Enero; 1955–1957) and Simon Rodriguez (1956–1957). Devoted to massive squatter relocation, these provided over 10,000 dwelling units.

16 Villanueva's El Silencio was among the first slum redevelopment projects to be supported by any Latin American government. By the 1950s however, its scale and cost was deemed an inappropriate response to the intensifying *rancho* settlements.

17 The goal of the New National Ideal was "the moral, intellectual, and material improvement of the inhabitants of Venezuela and the rational transformation of the physical environment." See Servicio Informativo Venezolano, *Venezuela bajo el Nuevo Ideal Nacional* (Caracas: Imprenta Nacional, 1954), frontispiece's facing page. This entailed the "war against the *rancho*" and the massive housing construction of the 1950s.

18 See Manuel López, "La Arquitectura del '2 de Diciembre,'" *Boletín del Centro de Investigaciones Históricas y Estéticas* 27 (1986): 148–172. To date, this remains the most detailed account of the *superbloque* venture.

19 Villanueva and Cepero, *La Vivienda*, unpaginated.

20 See Eric Mumford, *The CIAM Discourse on Urbanism, 1928–1960* (Cambridge: MIT Press, 2000), 190–191.

21 See Henry-Russell Hitchcock, *Latin American Architecture since 1945* (New York: MoMA, 1955), 137.

22 Ibid., 37.

23 Guido Guazzo, "Lettera da Caracas," *L'Architettura: Cronache e Storia* 17 (1957): 798–806.

24 André Bloc, "Urbanisme des Capitales," *L'Architecture d'aujourd'hui* 88 (1960): 1–2.

25 May Lumsden, "Caracas: the Buildingest City in South America," *Architectural Forum* 101 (1954): 152.

26 See *L'Architecture d'aujourd'hui* 74 (November 1957). The 23 de Enero complex is shown on the issue's cover.

27 José Matos Mar, *Urbanización y Barriadas en América del Sur* (Lima: Instituto de Estudios Peruanos, 1968), 294.

28 Juan José Martín Frechilla, *Diálogos reconstruidos para una historia de la Caracas moderna* (Caracas: UCV/CDCH, 2004), 99–152.

29 Eric Carlson, "Evaluation of Housing Projects and Programs: A Case Report from Venezuela," *Town Planning Review* 31, no. 3 (October 1960): 207.

30 Ibid., 201.

31 Ibid., 199.

32 Banco Obrero, *Proyecto de Evaluación de los Superbloques* (Caracas: Tipografia Vargas, 1959).

33 Richard Harris provides an overview of the consolidation of economic discourse and self-help within international housing policy in articles including: Harris, "The Silence of the Experts: 'Aided Self-Help Housing' 1939–1954," *Habitat International* 22 no. 2 (1998): 165–189; Harris and Godwin Arku, "Housing and Economic Development: The Evolution of an Idea since 1945," *Habitat International* 30 (2006): 1007–1017.

34 Carlson, "Evaluation," 207–208.

35 See in particular *Manual para la organización de proyectos piloto de ayuda propia y ayuda mútua en vivienda* (Bogotá: CINVA, 1953).

36 By 1962, Venezuela had more self-help dwelling units supported by the Inter-American Development Bank than any other country in Latin America. Likewise, Venezuela was the region's most active proponent of urban community development projects. See Marcia N. Koth et al., *Housing in Latin America* (Cambridge: MIT Press, 1965); Francis Violich and Juan Astica, *Community Development and the Urban Planning Process in Latin America* (Los Angeles: University of California Press, 1967).

37 See Arturo Almandoz, "Longing for Paris: The Europeanized Dream of Caracas Urbanism, 1870–1940," *Planning Perspectives* 14 (1999): 225–248.

38 For recent scholarship on the importance of low-income housing history within broader transnational discourses, see Leandro Benmergui, "The Alliance for Progress and Housing Policy in Rio de Janeiro and Buenos Aires in the 1960s," *Urban History* 36, no. 2 (2009): 303–326.

39 Charles Abrams, *Man's Struggle for Shelter in an Urbanizing World* (Cambridge: MIT Press, 1964), 52–53.

40 Josep Lluís Sert, Robert Moses, and Francis Violich were all called upon to consult on specific issues during the formulation of the 1951 Caracas Master Plan. For his critiques, see Francis Violich, "Caracas: Focus of the New Venezuela," in *World Capitals: Towards Guided Urbanization*, ed. H. Wentworth Elredge (New York: Anchor Press, 1975), 256.

41 Despite recommendations to prohibit *ranchos* on *superbloque* sites, a decade after their construction most open spaces had been occupied, as John Turner also deplored. For his initial reaction and later critique see, respectively, "Mass Urban Re-housing Problems: Superblock Program of Banco Obrero, Caracas, Venezuela, 1954–1958," *Architectural Design* 33, no. 8 (August 1963): 373–374; Turner, *Housing by People: Towards Autonomy in Building Environments* (London: Marion Boyars, 1976), 45.

42 D. J. Dwyer, *People and Housing in Third World Cities: Perspectives on the Problem of Spontaneous Settlements* (London; New York: Longman, 1975), 129.

43 Several housing specialists set Caracas as the antipodal example to Singapore and Hong Kong: whereas the two Asian cities illustrated how well-managed, vertical, high-density typologies could positively respond to the urgency of low-income dwelling provision, Caracas came to demonstrate the opposite condition. See Jaqueline Tyrwhitt, *High-rise Apartments and Urban Form* (Athens: Athens Centre for Ekistics, 1968); Dwyer, *People and Housing*.

44 Such disapproval provoked considerable efforts for the reappraisal of figures like Villanueva as crucial contributors to Latin American modern architecture despite their involvement with Pérez Jiménez's drastic housing policies. For ongoing efforts to assert Villanueva's contribution, see www.fundacionvillanueva.org.

45 Carlson, "Evaluation," 208.

46 The TABO's socio-political aims would only later be publicly stated. Guido Bermúdez, for example, referred to the once hoped-for result that the high-rises would increase collective conscience and engender the fall of the dictatorship. Cepero et al., "Conversación," 80.

47 See Juan Pedro Posani, "El rancho y el superbloque," in *Caracas a través de su Arquitectura* (Caracas: Fundación Fina Gómez, 1969), 526–532.

48 See for example the integrated rehabilitation of La Cañada *superbloques* within the 23 de Enero complex.

49 Bitter and Weber, *Caracas,* 3.

50 Project co-initiator Felix Madrazo specified three ethnographic findings that were considered relevant for the design: the distribution of families across both *rancho* and *superbloque*; the acknowledgement of the high-rises as valuable architectural artifacts; and a qualitative densification process allowing for the high accessibility of public facilities. These led to the museum's conception as "another layer of *barrio*/block integration." Felix Madrazo, personal communication with author, May 2011.

51 Ibid.

52 Bitter and Weber, *Caracas,* 14.

53 Mateo Pintó, "Urban Housing for the Twenty-first Century" (MSc thesis, FAU/UCV, 1999).

54 Alfredo Brillembourg and Hubert Klumpner, "Urban Acupuncture in Caracas: Bottom-up Strategies in Venezuela's Capital," *Topos* 64 (2008): 24–29.

55 For a discussion of the "orientalist" fascination for the self-built city, see Thomas Angotti, "Apartheid in Postcolonial Latin American Cities," paper presented at the Latin American Studies Association Congress, Rio de Janeiro (June 2009).

56 On this point see Ann Varley, "Postcolonizing Informality?" in *The Production, Use and Dissemination of Urban Knowledge in Cities of the South*, eds. V. Brunfaut, V. d'Auria et al., N-Aerus XI Conference Proceedings (Brussels, October 28–30, 2010), 305–317. See also Félipe Hernandez et al., eds., *Rethinking the Informal City* (Oxford: Bergahn, 2010).

57 "A *superbloque* is the best to live in, don't say there's anything better / because I'm friends with Julián and with Martín, and they'll see I'm cool here / Go find a guy from another place, I won't change you for my *superbloque*." "El Superbloque," song lyrics by Venezuelan singer and composer Simón Díaz.

58 Beatriz Jaguaribe, "Modernist Ruins: National Narratives and Architectural Forms," in *Alternative Modernities*, ed. Dilip Gaonkar (Durham: Duke University Press, 2001), 347.

59 For descriptions assuming the *superbloques* as a "modern failure" see Martin Pawley, *Architecture versus Housing* (New York; Washington: Praeger, 1971), 81.

60 On the impact of foreign influence on Venezuelan architectural identity, see Azier Calvo Albizu, *Venezuela y el problema de su identidad arquitectónica* (Caracas: UCV/CDCH, 2007).

61 Cepero et al., "Conversación," 80.

62 The "Progressive Development Units" launched in Ciudad Guayana exemplified this model. See Carlos Reimers, *After Sites and Services: Planned Progressive Development Strategies in Low-Income Housing during the 1990s* (MSc thesis, McGill University, 1995).

63 See Ronaldo Ramirez et al., "The Commodification of Self-Help Housing and State Intervention: Household Experiences in the *Barrios* of Caracas," in *Beyond Self-Help Housing*, ed. Kosta Mathéy (London; New York: Mansell, 1991), 95–144.

64 Carlson, "Evaluation," 204–207.

6

MONUMENTALITY AND RESIGNIFICATION

The UNCTAD III Building in Chile[1]

Daniel Talesnik

The construction in Santiago of the United Nations Conference on Trade and Development (UNCTAD) building, consisting of a low-rise and tower complex, speaks to a short-lived spirit that prevailed in a section of Chilean society during Salvador Allende's presidency, representing a peak of the period's idealism. The building did not engage with propagandistic socialist aesthetics, yet the political aims of the period are manifest in the construction process—the empowerment of the workers, and the deployment of technology to meet a tight timeline—and ultimately in the building's performativity and open relationship to the city. With the coup d'état in September 1973, the Military Junta headed by General Augusto Pinochet took hold of the complex and significantly altered its use, curtailing the civic purposes of the building. What was conceivably the most paradigmatic building of Allende's presidency instead became emblematic of the military government, in diametrical opposition to its original identity. In this chapter, I examine the political trajectory of the UNCTAD building, particularly the low-rise, through three periods of its existence—the Unidad Popular government (1970–1973), the military dictatorship (1973–1990), and the return to democracy (1990–present)—and analyze the design and construction from material, technological, urban, artistic, and social perspectives. This analysis of the UNCTAD building leads to a questioning of the relationship between architecture and politics, enabling speculations about monumentality and resignification in architecture.

The story of the building begins in 1971, when, framed by the optimism of the early days of his tenure, Allende petitioned the United Nations to host the UNCTAD III conference scheduled for May 1972. Established in 1964, UNCTAD's mission is to facilitate the integration of developing countries into the world economy, holding conferences every four years. In 1970, just a year earlier, Salvador Allende had become president of Chile and the first democratically elected Marxist leader in the Americas. Chile entered a period of political experimentation during which the fervor of Allende's supporters was met with increasing unrest from his detractors.[2] From a global perspective, Chile became an unlikely theater of the Cold War: if Allende's experiment was successful, it could set a precedent for other countries. The engagement with a UN agency conference was seen by Allende's government as an opportunity to showcase the country and to win over the skeptics of the "Chilean road to Socialism."

Chile lacked the facilities for an international event like an UNCTAD conference, so Allende asked the Chilean Engineers and Architects Associations to evaluate the possibility of building a new conference center. The two and a half years that they estimated as necessary to search for a site, outline a program, hold a competition, develop the project, and solicit construction bids, made the May

6.01 View toward the
northwest of the UNCTAD
building under construction,
1971

1972 deadline impossible to meet. Allende turned to the Urban Development Corporation (CORMU), the state agency that was to manage the project, and asked for an alternative assessment. The head executives of CORMU, the architects Miguel Lawner and Jorge Wong, assured Allende that a new building would be possible in less time. CORMU counted on its legal powers, on the dynamic coordination of diverse public agencies, on assistance from the recently nationalized industries, and on the commitment among Allende's supporters to have the building delivered on time. Allende ordered all government institutions to fast-track the legal procedures, and named a special assessor to negotiate the budget approval with the congress.[3] A project advisory committee was created, and architects José Covacevich, Juan Echeñique, Hugo Gaggero, Sergio González, and José Medina were selected to lead the design, with Lawner as coordinator of the overall project on behalf of CORMU. The architects had not worked together in the past, and not all were "militants" of the president's national project, yet they demonstrated a profound commitment to the task Allende had given them. There was an architectural problem to solve, and it was to be answered with assertive solutions. The new building was designed, built, and inaugurated in 275 days.

The building had to fulfill the needs of the UNCTAD conference and accommodate its subsequent conversion into a cultural center. An appropriate site had to be chosen, and the final decision was a 656 by 1640 ft. lot in the northern portion of the ongoing San Borja urban renewal project, which comprised several city blocks that were being renovated with twelve new residential towers. The site was in the first blocks of the Alameda—the main avenue in Santiago—in the vicinity of the Parque Forestal, Museum of Fine Arts, National Library, Municipal Theatre, and one of Santiago's main universities.[4] Since the zone was currently undergoing renovation, the insertion of a new building into the urban grid would be a less complicated affair; moreover, a station for the new subway system had already been planned next to the site. Finally, the chosen site belonged almost entirely to CORMU, and immediately behind it one of the new residential towers was already being built, and could be transformed into an office tower with no major structural alterations. This left the low-rise building as the primary design and construction challenge.

6.02 Aerial view of the UNCTAD building construction site, with the Secretariat tower, and four of the San Borja refurbishment towers on the other side of Alameda Avenue, Santiago, 1971

José Medina traveled to New York to develop the program with the UN's chief architect. The architectural brief, loosely based on the UN headquarters in New York, included a 2,300-seat plenary assembly room, several large conference rooms, as well as two dining areas seating 600 and 200 people respectively. Shops, bank branches, travel agencies, a post office, communications room, delegates' halls, and a myriad of minor services were also distributed throughout the four-story low-rise structure—a reinforced concrete and prefabricated metal building housed under a 97,000 sq. ft. metal roof supported by sixteen oversized reinforced concrete pillars. The total program required 260,000 sq. ft., half of this being office space for international delegates, to be housed in the twenty-two-story tower behind the site.[5] The low-rise and tower—a combination utilized in existing UN organizational buildings, such as UNESCO in Paris and the UN headquarters in New York—were connected by service bridges on three levels to form the UNCTAD complex.

In order to meet the deadline, the architects and structural engineer Carlos Sandor decided to speed up construction by employing two independent structural systems. The so-called "circus tent" strategy allowed construction of the superstructure to begin before there was a final design for the interior. The only way of advancing simultaneously with the roofing and the upper floors was a hybrid building topped with a metal structure assembled using extramural cranes. Special mention should be given to the explicit objective of exploiting Chile's productive and technological capacity in the building. Due to the fragile domestic economic situation and the credit blockade erected by the United States against Allende's government, a lack of foreign currency made it difficult to import building materials. By the end of 1971, the government had nationalized all major mining firms and a number of other companies, so the majority of the construction materials, such as Corten steel, a novelty in Chile at the time, came from nationalized businesses or companies associated with the government.[6] The focus on national production to serve local purposes reflected the government's aims of shifting from a dependent to a "free" economy.[7] However, Lawner and Medina

6.03 Plenary assembly
room, UNCTAD building

6.04 Northeast side
of the UNCTAD building
construction site,
Santiago, 1971

affirm that the use of Chilean steel for the roof had no symbolic meaning or
resonance with prevailing political discourse—so a "steel-socialism" connection
would be misleading. According to Medina, the Corten steel was primarily used
for its structural characteristics and "zero maintenance."[8] Nonetheless, the fact
that the high-rise was crowned with a band of Corten steel that maintained the
visual unity of the complex was also a symbolic move.[9] Thus the use of local
materials became both a practical and a political decision.

The fast construction was made possible by three non-stop shifts per day that
were coordinated by state-of-the-art computational software. The use of
computerized technology enabled efficient scheduling and monitoring progress
in great detail on a daily basis. Chilean engineer Helmut Stuven introduced the
Pert software, which had been created by the US Navy in 1958 for the fabrication
of Polaris missiles, and was later used for the design and construction of Apollo
11. The Pert software analyzed all data and details at every scale of the project,
from the bolts, to the temperature, and even the general mood of the workers—
anything could be a variable—and produced a Pert chart or "critical diagram" of
hundreds of activities, as well as auxiliary charts with minor tasks. A significant
innovation in the UNCTAD building was that instead of having one person
in charge of data collection, Stuven taught seventy-five construction workers
how to feed data into the IBM software cards—a fact that speaks to the
unprecedented responsibilities given to the workers during the building process.
Every Wednesday, a print-out of the chart was delivered to CORMU in order to
evaluate the development of the construction.[9] This pioneering coordination of
the construction speaks to broader government initiatives aimed at improving
the efficiency of national production. The UNCTAD construction process served

Allende as a model of what, given the appropriate circumstances, Chilean industry could achieve.

The compressed timeline foregrounded the role of the workers who made the prompt construction of the building possible. Tradesmen were highlighted in the press alongside Allende, who spoke of them as the incarnation of the ideals of his government, echoing the frequently used slogans of those days: "government of the workers" and "workers into power." In the process, the construction worker was used to publicize the triumph of the Chilean people. Since a collective of architects—not a single architect—designed the building, it allowed for a shift in focus from the sole genius as intellectual author, to the cooperative effort of the workers as the material authors of the building. One incident illustrating this is that halfway through construction, a strike by professionals prevented engineers and architects from visiting the site to supervise the work. Nonetheless, during that time one of the construction managers took charge of the project, and the workers kept on schedule.[10] The workers were understood as both builders and clients: since the building was to become a cultural center after the conference, it was ultimately by the people and for the people. The construction of the building also integrated "the people" at a wider societal level, as organized crews of students helped in the construction on the final weekends, and members of the Central Union of Workers voluntarily embellished the surroundings.

The most prominent local artists and artisans were included in the design process, not only in the creation of close to forty artworks that were distributed throughout the building—thereby fulfilling the need for an exhibition space of Chilean contemporary art—but also in the design of the lighting, furnishing, ventilation, and acoustics of the building. The design department of a state institution led by the German industrial designer Gui Bonsiepe created the signage.[11] Every artist was paid the equivalent of the monthly salary of a specialized construction worker.[12] This equalizing of artists and workers spoke of the attempt to create a communion between "high" and "low" art, both being equally important for Allende's discourse.

According to the project report produced during the construction, the UNCTAD low-rise would be scaled for functions of collective use, the neighborhood, and the city. The architects aimed to reinforce the linearity of Alameda Avenue through the marked horizontal design of the low-rise, and to link the northern and southern parts of the neighborhood through a subway station tunnel.[13] Moreover, the building itself functioned as a pedestrian passageway by connecting Alameda Avenue with the Villavicencio neighborhood and the Parque Forestal north of the complex. For the architects, the design of the ground floor of the building, providing multiple points of access, emulated a plaza that could accommodate crowds of people. The gallery-like sidewalk was intended to evoke the traditional porosity of colonial galleries—a contestable point, however, since the extreme height of the roof diffused this intention. The architects aimed at modifiable "flexible" spaces within the building, conceiving the "marrow" or white cubes that acted as an envelope on the top floor of the low-rise as a provisional solution that could be modified in the future. The interior sections were "removable" and the "table scheme" superstructure was seen as a solution that allowed for the altering of interior elements. What is evident—this being one of the main points for which it was criticized in later years—is that the building was boldly implanted into the urban grid of Santiago, using its size and rough aesthetic to impose an unprece- dented type of architecture in the city center. The UNCTAD building, when finished in 1972, intersected with a series of current architectural debates: stylistically, the low-rise can be described as a megastructure with Brutalist undertones, where the bareness of some of the main materials accentuates its massiveness.[14]

6.05 Carpenter Eulogio Maldonado hands over the building to President Salvador Allende

6.06 (above) Construction workers at the UNCTAD building, Santiago, 1971

6.07 (below) Conference room, mural by Eduardo Vilches

6.08 (above) Dining hall, wickerwork fish by Manzanito

6.09 (below) Delegates' room, lighting project by the architects and Bernardo Trumper, artwork on the walls by Mario Toral

6.10 (above) Detailed cross-section, José Covacevich, Juan Echeñique, Hugo Gaggero, Sergio González, and José Medina, UNCTAD building, Santiago

6.11 (below) Night view toward the northeast of the UNCTAD building, ca. 1972

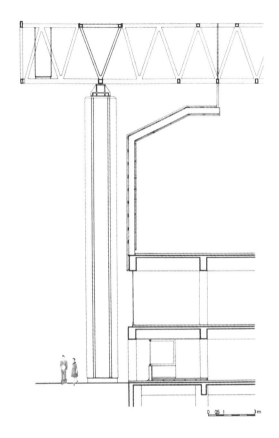

When inaugurated, the UNCTAD building was a matter of national pride, and was praised within the architectural community. In order to evaluate the relevance of the UNCTAD building, a comparison with the Economic Commission for Latin America (CEPAL) building (1960–1966) by Emilio Duhart is useful. As an institution, the original UNCTAD program was similar to that of the CEPAL, but scaled up to a global level.[15] The CEPAL made Santiago its headquarters in the beginning of the 1960s, turning the city into a regional center for the UN—this in turn favored granting the UNCTAD conference to Santiago.[16] The CEPAL is located in the middle of a large site by the Mapocho River, amid a residential neighborhood in the northeast part of the city, on an enclosed and guarded terrain. Its urban isolation led architectural historians Roberto Segre and Rafael López Rangel to criticize the enclosed nature of the building and the autonomy of its symbolic system. They argued that an international organization such as CEPAL, which works on the continent's social and economic problems, should be located in an urban area open to the contact and participation of the community. Ultimately, it is important to note that the UNCTAD building's interaction with the city—with its unconcealed relationship to the neighborhood where it was located—operated in diametrical opposition to the CEPAL. It is through this performativity that the UNCTAD building can claim to have engaged with the political agenda of Allende—in its program, relation to the street, and openness to the community.

After the conference, the UNCTAD building was placed under the administration of the Ministry of Education and renamed the Gabriela Mistral Cultural Center. The building was among other things to show the art collection of the Museo de la Solidaridad, and Allende highlighted its open-ended cultural program, declaring in a public speech that it was to become "the material base of the great Institute of National Culture."[17] The first and only year that the cultural center functioned was the last year of Allende's government, which saw great upheavals—especially in the months that followed the parliamentary elections of March 1973, which led to an increasing polarization of Chilean society. After the conference, in December 1972, the left-wing magazine *La Quinta Rueda* published an article entitled "Living UNCTAD Style." The article addressed the socio-political dimension of the building some months into its use as a cultural center, underscoring an "illusion of modernity," both in the surroundings and the interior of the building. In the same rooms where the international delegates had held meetings, now there were miscellaneous shows, or meetings held by trade unions and other organizations. One of the main reasons to visit the building was the desire to experience the new social equality promoted by the government, performed in the "self-service" cafeteria, which could cater to 600 customers simultaneously and was serving an average of 1,500 subsidized lunches per day.[18] This was a source of special pride for the government, and the cafeteria model was meant to be replicated in other major urban centers. In the cafeteria, students ate side by side with construction workers. The article made reference to this incipient aspect of Chilean society: the building became a gathering point, a pole of urban attraction, where one could suppose that social differences could be set aside. Social equality had a theater in this building, a stage for a daily performance of an idealized country.

The UNCTAD building was at the frontier of Allende's populism and promotion of social equality in other more nuanced ways. Allende famously told the workers that they did not need to "dress up" when visiting the building, that they could visit in their "battle attire," in clear allusion to their "work garments." Toward the end of the article, however, a note of skepticism contradicted Allende's hopes. "The only thing missing from the scene," the journalist argued, are people speaking English, French, or Esperanto. The style is international; nevertheless, what can appear to be "cool" to a student is still foreign to construction workers, those that the president declared should feel themselves to be the masters of the building. Living UNCTAD style smells little like Chile and a lot like the movies, it still intimidates workers.[19]

6.12 View of the self-service cafeteria

6.13 Dr. Salvador Allende, President of the Republic of Chile, has lunch in the UNCTAD self-service cafeteria amid a student meeting

This uneasiness was reinforced by a worker's testimony:

> It makes me proud, comrade, that in Chile we build luxurious things, and workers do not have to show an ID at the entrance. But one cannot feel like the owner of this huge mass, because one arrives home at night to a cabin in a shantytown, and it has always been like this.[20]

These comments express the sense of urgency of the period, particularly for the less privileged.

The social dimension of the building, its cultural and communal phase, lasted only a year. On September 11, 1973, the Chilean army, headed by General Pinochet—with the full support of the Chilean right, the Christian Democrats, and backed up by the United States—staged a coup in which Allende reportedly committed suicide at the Chilean presidential palace, La Moneda, while it was being bombed. Due to the damage inflicted on the nation's main administration building, leaving it uninhabitable, the new military regime hastily moved its command center to the UNCTAD complex located only thirteen blocks east on the same avenue. The building was renamed from "Gabriela Mistral," a unifying national figure and 1945 Nobel Prize laureate in literature, to "Diego Portales," a polemical and divisive nineteenth-century conservative—a harbinger of the building's future under the military. The junta used the low-rise for the offices of the executive and what stood in for the legislative power (since the congress was dissolved throughout the dictatorship), and the tower for the Ministry of Defense.

The building's central location in Santiago, its adaptable program, its monumental character, and symbolic importance led the military to select it as an administrative

6.14 Entrance foyer

center. Due to security measures, described by the military as an "armoring" of the building, the original transparent relationship with Alameda Avenue through the windowed facade at the plinth level was blocked with masonry. The perimeter was fenced, and armed forces stood on permanent guard; overall, the building was isolated both visually and physically from its surroundings. The lateral door on Alameda Avenue designed by artist Juan Egenau, which had allowed direct access to the two theaters was now sealed. The cafeteria was closed, and all public circulation through the building was banned, radically changing its civic character. On the main wall of the plenary conference room, a copper plate with the words "1810–Chile–1973" was installed, and on the stage stood an oversized podium with a Chilean shield. The overt connection between Chile and copper—the country's main natural resource—and the deployment of nationalistic dates and symbols completely transformed the character of the interior. Such physical changes embodied what Luis Hernán Errázuriz has described as the "aesthetic-cultural coup d'etat" carried out by the Pinochet regime, which entailed visual, aural, and spatio-cultural aspects. Errázuriz points out that what may appear as incidental changes to the building have deep political meaning, such as painting over the ventilation system—originally in bright red by artist Félix Maluenda—with light "army" green.[21] Errázuriz further develops the concept of symbolic violence, stating that, "the military regime imposed through symbolic acts both the dominion and exercise of power and the eradication of the ideas of the defeated government."[22] Another example was the looting of the building's art pieces, of which to date only fourteen have been retrieved. The military carried out a nationwide "sanitization" of all elements related to Allende's government, and some of the transformations in the UNCTAD building directly relate to this campaign.[23]

These measures allowed the building to be used once more, this time strictly for political events, where Pinochet held court over the junta and the country. In March 1975, economist Milton Friedman delivered a lecture in what was now the Diego Portales building, titled "Chile and Its Economic Take-Off" as part of an international conference that aimed to correct the previous years of anti-market policies.[24] In June 1976, US Secretary of State Henry Kissinger attended the general conference of the Organization of American States held in the building. The memorandum on Kissinger's conversation with Pinochet notes that he found the building beautiful—before the discussion turned to topics which (in hindsight) reflect the complicity of the United States in Chilean politics of this period.[25] Kissinger's comments suggest how successful the sanitization had been, to the point that it maintained no association with Allende's government, and that apparently he was not briefed about the building's history.

In 1981, the executive power returned to La Moneda. The Diego Portales low-rise building remained under the control of the Ministry of Internal Affairs and the tower under the Ministry of Defense. In 1988, the building acted as the press center for the national plebiscite, in which Chileans voted "No" to the continuation of Pinochet, putting an end to his dictatorship—adding yet another layer of meaning to the building in Chilean collective memory. After Pinochet's defeat, the country resumed its democratic tradition with the 1989 elections. As of 1990, the Diego Portales complex remained under the Ministry of Defense, used as a public and private convention center and as the press center for Chilean elections. Since it was the only building with this type of facilities in the city center, renting it was a highly profitable operation. Successive democratic governments did not reappropriate the building for civic uses and its name remained unchanged.

In *Non-Places*, Marc Augé analyzes the relationship of spatial forms and symbols, speaking specifically about buildings that house sovereign rulers, and how the identification of power with the place from which it is exercised is a political

discourse inherent in modern states. Augé makes the point that metonymy allows us to speak of a country by mentioning its capital, and a capital by the building that houses the ruler: "the White House" or "the Kremlin" speak in parallel about a monumental place, an individual, and a power structure.[26] One could conclude that since Pinochet exercised his power from the UNCTAD building during the initial eight years of his dictatorship—this being the most repressive period of his regime—it was this meaning that became fixed to the building in Chilean memory, overshadowing its previous uses as a UN conference building and as a cultural center. Although afterward Pinochet did rule from La Moneda, once his regime ended, the Chilean presidential palace necessarily returned to its traditional role as a synonym and symbol of democracy.

The original UNCTAD building can be linked to what Sigfried Giedion, Fernand Léger, and Josep Lluís Sert described in their "Nine Points on Monumentality" as a constructed and emotional quality of buildings rather than sheer structural bigness, a monumentality in dialogue with the user/viewer: "The people want the buildings that represent their social and community life to give more than functional fullfilment. They want their aspiration for monumentality, joy, pride, and excitement to be satisfied."[27] This idealized representation of society was present in the design, construction, and program of the UNCTAD building. Through the participatory construction process, and its day-to-day use in the last year of Allende's government, this building orchestrated a sense of pride in Allende's supporters around his political project. All these impulses reinforced the physical monumentality of the building: particularly in its iteration as a cultural center, the building engaged with social monumentality through the strongly associated meanings that gave it relevance.

The military transformed this monumentality of the building through physical alterations that contradicted the building's use under Allende. By resignifying the

6.15 General Augusto Pinochet leaving the Diego Portales building, 1988 building, the military erased its previous significance, and its journey from symbol of a radical left to symbol of a military right reveals the ambiguities inherent in monumentality itself. A "big building" can serve different masters; social monumentality, on the other hand, has to be collectively tested. The successive changes of identity intensified its symbolic dimension to the point that once Chile was again a democracy, the first three governments did not know what to do with such a charged architectural icon, and thus left it functioning as a convention center, to its own detriment.

On March 5, 2006, a fire caused by a short circuit—related to the lack of maintenance—destroyed a large part of the low-rise building. On the day of the fire, a crowd gathered to watch the conflagration, and when the roof-structure collapsed, they cheered.[28] This episode speaks clearly to the contested meaning of the building: the aftermath of the fire inaugurated a public evaluation, where detractors spoke of a building that had been "premature" in its design and oversized for its urban context and argued for its demolition.[29] Ultimately, and with the help of the Chilean Architects Association, the building was officially declared public property as part of Chile's national heritage. The building deserved to be preserved. The surviving concrete pillars were left in place, and the damaged portion of the roof was removed. The National Council of Culture took over the administration of the building, and in 2007, a public competition for its reconstruction was held. Today a refurbished building—re-renamed Gabriela Mistral after thirty-four years—doubles as a library and cultural center.[30]

Despite the efforts made after the fire, in a global twenty-first-century Santiago the refurbished/reconstructed cultural center—clad in perforated plates of Corten

6.16 View toward the northwest of the Gabriela Mistral Cultural Center, Cristián Fernández Arquitectos, Lateral Arquitectura & Diseño, 2009–2010

steel that partially hide some of the perimeter pillars—does not achieve the relevance that the building had in 1972. Today, both the steel cladding and the cultural center typology are ubiquitous, if not generic. The socio-political relationship between the residents of Santiago and the building is hardly comparable to the experimental symbiosis of user and building that was beginning to emerge in the 1972 cultural center; neither does the building retain the level of significance it had, for totally opposed reasons, under the military. After returning to democracy, the building was not re-charged with meaning, and its significance and potential monumental value were lost. In an important way, the building had already ceased to be resignifiable and monumental even before the disastrous blaze.

Notes

1   My thanks to Miguel Lawner and David Maulén for their help during the research. Unless otherwise noted, all translations are by the author.
2   Allende's Unidad Popular coalition (comprising Communists, Socialists, "Radicales," and a number of Christian Democrats) won the most votes, but without an absolute majority the parliament had to confirm the validity of the elections.
3   Jorge Wong, "La Maratónica Construcción del Edificio UNCTAD III," July 21, 2008, http://www.consejodelacultura.cl.
4   Miguel Lawner, "UNCTAD III Centro Cultural Metropolitano de Santiago," *AUCA* 22 (1972), 58–59.
5   The central nucleus of the tower was modified to increase the number of elevators, and a cantilever was added to the perimeter of each floor to augment the floor space. The exterior was covered with a *brise-soleil* that reinforced its verticality, and differentiated it from the residential towers in the San Borja project.
6   The nationalization of industries, instrumental for the rapid construction of the UNCTAD building, was one of the most controversial initiatives of Allende's government and ultimately among the reasons the military claimed for the coup.
7   See Eden Medina, "Designing Freedom, Regulating a Nation: Socialist Cybernetics in Allende's Chile," *Journal of Latin American Studies* 38, no. 3 (August 2006): 579; and Richard E. Feinberg, "Dependency and the Defeat of Allende," *Latin American Perspectives* 1, no. 2 (Summer 1974): 30–43.
8   José Medina, e-mail message to author, December 19, 2008.
9   I am indebted to David Maulén for this information.
10 David Maulén, "El Edificio del Pueblo Antes de las Llamas," *La Nación*, March 12, 2006.
11 Miguel Lawner and David Maulén, "UNCTAD III, Santiago de Chile, 06/1071–04/1972," *Documenta Magazine* 1–3 (2007).
12 Francisco Brugnoli, "Habitares: Espacios Posibles e Imposibles," *Simposio Internacional Utopías* (Santiago: Ministerio de Educación, 1993), 96.
13 The subway station was not finished in time for the conference, but during the Pinochet era, and so was never connected to the building.
14 Lawner mentions that some of the architects were interested in Kunio Maekawa's Tokyo Metropolitan Hall (1958–1966), whose influence can be observed in the form and finishing of the concrete pillars. Medina points to another megastructure that deserves special attention: Roche, Dinkeloo, and Associates' Veterans Memorial Coliseum and Knights of Columbus tower in New Haven, Connecticut (1968–1970); Medina had worked for Roche in the late 1960s, hence his awareness of this project. Miguel Lawner, e-mail message to author, December 17, 2008; José Medina, e-mail message to author, December 19, 2008.
15 See Joseph Love, "Latin America, UNCTAD, and the Postwar Trading System," in *Economic Development in Latin America: Essays in Honor of Werner Baer* (Houndmills: Palgrave Macmillan, 2010), 22–33.
16 Roberto Segre and Rafael López Rangel, *Architetture e territorio nell'America Latina* (Milan: Electa Editrice, 1982), 109.
17 The Museo de la Solidaridad was an initiative of foreign intellectuals: the aim was to collect works donated by artists from all over the world in support of Allende. Between 1971 and 1973, 400 artworks were donated. Miguel Lawner, e-mail message to author,

October 1, 2011; Salvador Allende, public speech at the inaguration of the Museo de la Solidaridad, May 17, 1972, *Chile: Breve Imaginería Política 1970–1973*, http://www.abacq.net/imagineria/discur8.htm.

18   Luisa Ulibarri, "Vivir Estilo UNCTAD," *La Quinta Rueda* (December 1972): 5.

19   Ibid.

20   Ibid.

21   Luis Hernán Errázuriz, "Dictadura Militar en Chile, Antecedentes del Golpe Estético-Cultural," *Latin American Research Review* 44, no. 2 (2009): 136–157.

22   Ibid, 137. Similarly, other architectural landmarks in Santiago such as the National Stadium and the Chile Stadium were transformed into provisional detention and torture centers in the days following the coup.

23   The sanitization extended to some of the architects involved in the UNCTAD building. Lawner was taken prisoner the day after the coup from the CORMU offices along with forty-three of his employees, and was eventually taken to the Dawson Island prison camp. Besides heading the CORMU, Lawner was in charge of the CORMU-Chilean Army Alliance, building housing for army personnel, which had made him extremely visible to the military—according to Lawner this was the main reason for his detention. Lawner was exiled from Chile in June 1975, finding asylum in Denmark, and returned to Chile upon authorization from Pinochet's government in March 1984. Miguel Lawner, e-mail message to author, September 30, 2011. Lawner is interviewed about his experiences under the dictatorship in Patricio Guzman's documentary *Nostalgia for the Light* (2010).

24   Friedman's former Chilean students, dubbed the "Chicago Boys," had a strong influence on Chilean economic policy under Pinochet. See Orlando Letelier, "The Chicago Boys in Chile: Economic Freedom's Awful Toll," *The Nation*, August 28, 1976; and Naomi Klein, *The Shock Doctrine* (New York: Metropolitan Books Henry Holt, 2007), 7, 12, 330.

25   US Department of State, Memorandum of Conversation between Augusto Pinochet, Henry Kissinger et al. June 8, 1976. Gerald Ford Library, Ann Arbor, MI.

26   Marc Augé, *Non-Places* (London; New York: Verso, 2008), 51–52.

27   Josep Lluís Sert, Fernand Léger, and Sigfried Gideon, "Nine Points on Monumentality," in *Architecture Culture 1943–1968*, ed. Joan Ockman (1943; New York: Rizzoli, 1993), 29.

28   I am relying on Miguel Lawner for this anecdote, who in turn received the information from a fireman on duty the day of the fire. This episode speaks of the change of meaning the building had undergone: the crowd, most likely associating the building with its use under the military regime, celebrated its destruction.

29   Javier Rojahelis, "¿Otro centro cultural deficitario?: Reconstruyendo desde las cenizas el Diego Portales," *El Mercurio*, August 12, 2007.

30   Cristián Fernández Arquitectos won the competition. The program includes a library, documentation center, conference rooms, and a cafeteria—a new theatre is still unfinished. Before the fire there had been two independent attempts to promote the building's refurbishment—one in 1998 led by the Instituto Río Colorado through a design proposal, and another in 2002 by the Chilean Architects Association through a design competition.

7
A PANEL'S TALE

The Soviet I-464 System and the Politics of Assemblage[1]

Pedro Ignacio Alonso and Hugo Palmarola Sagredo

The click of the camera's shutter captured the moment. On November 22, 1972, President Salvador Allende is shown surrounded by hundreds of Chilean workers and Soviet dignitaries in a new factory in the small industrial town of El Belloto, Quilpué, northwest of Santiago. Standing at the center of the photograph Allende bends down to inscribe his signature into the wet cement of a panel—one of the first produced by the factory's assembly line. For posterity, alongside his name he also writes an appreciative greeting to "Soviet and Chilean comrades," thanking his countrymen who would run the plant, and the country that donated the entire facility. Against this backdrop, of a small town on the edge of a country itself seemingly on the far edge of the world, Allende's inscription transformed a standardized industrial component into a unique monument, a symbol of his express hope and mission to carry out the economic and social transformation of Chile toward an "integral, scientific, and Marxist socialism."[2]

7.01 President Salvador Allende signs a commemorative KPD panel during the inauguration of the plant, November 22, 1972

As one of the few Soviet contributions to the "Chilean road to socialism," this factory was designed to manufacture prefabricated concrete panels for Chile's then fledgling program of social housing—a system that went by the name KPD after the Russian acronym КПД, meaning "large panel construction" (крупнопанельное домостроение).[3] Known in the Soviet Union as the I-464 system, this was an adaptation of the French Camus system, originally patented in 1948 to prefabricate large concrete panels in factories capable of producing up to 2,000 housing units a year.[4] Camus and I-464 were both manufactured on the basis of concrete panels held together via steel rods, with the joints between each panel filled with poured concrete to produce a rigid and homogeneous whole. Among the improvements to the system made by Soviet engineers, however, were the introduction of a linear cast concrete production process and adaptations made to the standard assembly plant, redesigned in the Soviet Union.

Despite a thawing of Soviet policy toward developing countries initiated in 1964 by Leonid Brezhnev, during the Cold War the only Latin American countries that were given KPD factories were Cuba in the 1960s and Chile in the 1970s. In both cases, the ability to produce affordable mass housing was desperately needed, partly as a result of recent devastating natural disasters—Hurricane Flora had decimated the eastern coast of Cuba in 1963, and the 1971 Chilean earthquake similarly destroyed many of the country's towns and much of the housing stock.

The Soviet donation of KPD and the introduction of an advanced housing system constituted a ready-made political program for a leader with the courage and savvy to grasp it. But the clear precedent for this lay in the Soviet Union itself. In the period immediately after Stalin's death, the country's gargantuan housing industry emerged as a fundamental component in the struggle for supreme political power.[5] Stalin himself had made the building industry one of the most important areas in the program for the socialist transformation of the national economy through his first five-year plan (1928–1933), but it was during the succeeding administration, led by Nikita Khrushchev (1953–1964), that all previous experiences in the production of industrialized housing were increased to an unprecedented economic and urban dimension. Under his government huge research grants were channeled into any technology that promised the acceleration of housing production, not just in the Soviet capital but also across

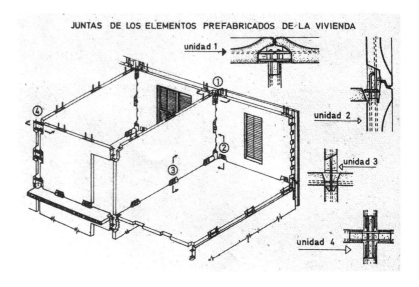

7.02 Axonometric scheme of the I-464 system in Cuba

7.03 Fidel Castro supervises rescue operations during the flooding caused by Hurricane Flora. *Ciclón*, directed by Santiago Álvarez, 1963

the most distant regions of its territory. The rapid dissemination of the resulting factories was an integral part of the regime's control and colonization strategies, deploying technological components as what Alexander D'Hooghe has termed "civilizing devices." In Cuba and Chile, as in the Soviet Union, the productive complex of KPD factories would be the most tangible territorial footprint of this logic, signaling that "the organization of space would be subordinated completely to the purpose of increasing production."[6]

On October 22, 1963, Cuban commander in chief Fidel Castro is seen standing on top of a tank struggling to move against the devastating floods caused by Hurricane Flora. The image—in Santiago Álvarez's documentary *Ciclón*—heroizes Castro's efforts against the troubled waters, as he seeks "to have a precise idea of the magnitude of the disaster, to directly command the tasks ahead, and to inspire soldiers and citizens with his example."[7] On its six-day journey, Flora killed about 1,200 people and left several hundred thousand homeless. Yet facing the tragedy, Castro was certain that the "Revolution is a force more powerful than nature."[8]

The Cuban engineer Vitervo O'Reilly recalled that Hurricane Flora came at the beginning of the revolutionary process, when several projects were underway to develop the country. But, "suddenly a weather event like this came to take the Cuban state a little out of its direction and planned actions." Direct partnerships with the Soviet Union were already in place: at the moment the hurricane struck the island, O'Reilly was in Moscow taking a course on prefabrication. It was then that "at our Embassy in Moscow *Ciclón* was screened for Nikita Khrushchev and his daughter"; moved by its dramatic scenes, "Nikita, in a gesture of solidarity with Cuba, decided to donate a factory to produce large concrete panel buildings."[9] A year after the Cuban Missile Crisis, this was a critical time for the two countries. A series of trade agreements were signed, aiming to rebuild mutual confidence; Khrushchev's gesture should be seen in light of this renewed spirit of cooperation.

Once in Cuba, the system—and the factory itself—was named the "Large Soviet Panel." By chance, only weeks before Flora reached Cuba, Castro had given a speech at the VII Congress of the International Union of Architects held in Havana. He questioned the meaning of industrialization in a country where three-quarters of the population were illiterate, but concluded that, "in order to begin solving the housing problem . . . we must first begin by solving the problem of the construction industry."[10] The factory was built under the direction of Soviet technicians in the city of Santiago de Cuba between 1963 and 1965. Beforehand a Cuban committee had traveled to Moscow to study adaptations to the design of the panels and buildings to meet the needs of the island. As O´Reilly observed, in the USSR construction "had to take into account the problem of snow loads, whereas in Cuba we had to take into account the wind load."[11] One of the most important modifications was made by architect Hugo D'Acosta: thinning and perforating the facade panels with a curved design that broke with the Soviet model. This achieved significant savings in concrete, allowed cross ventilation, and created a less monotonous expression. The Cuban transformations thus conceived the panels as a lighter, sculptural piece, bringing together the local demands for technical efficiency with a more "plastic" appeal. The urban planning of the first project, installed near the factory, was headed by the architect

7.04 Production process in Cuban I-464 plant

7.05 Interior view of I-464 building in Cuba

Edmundo Azze, and was based on the Soviet model of zoning according to districts and micro-districts.[12] The first residents were former shanty dwellers, those affected by the cyclone, as well as factory workers.[13]

The program entailed some political strife. Victor Rodríguez, a former factory worker, recalled that the United States had recently shelled a newly nationalized TEXACO oil refinery, also located in Santiago de Cuba. Although Rodríguez recognized that the panel factory was a lesser strategic target:

> We already had a socialist revolution, and the revolutionary fervor made us alert because at any moment the Americans would attack or sabotage the factory. . . . It was a Soviet factory, booming with revolutionary spirit . . . it could have been sabotaged just for being Soviet.[14]

Even though such sabotage never occurred, someone once "put counter-revolutionary slogans on one of the panels in the courtyard. There were writings against Cuba, painted on a panel."[15]The factory's yard became a sort of canvas for opposing ideological slogans, at other times in favor of the revolution. A poster was displayed close to the panels: "Nothing will stop the liberation of peoples."[16]

The frequent assumption that in Chile, as in Cuba, the donation of large panel factories was purely a mechanism to disseminate Soviet ideals needs to be qualified within the specific policies toward Chile overseen by Khrushchev's successor, Leonid Brezhnev. Although the Soviets expressed support for governments considered both socialist and anti-imperialist, by the end of the 1960s, and perhaps due to the Cuban and Chinese experiences in the Cold War, the relationship between the USSR and Chile can be characterized as rather cautious and pragmatic.[17] This, of course, was a period when the Marxist Unidad Popular led by Allende was increasingly forced to rely on the Soviet Union for financial support due to the US embargo. But documents from the conservative Christian Democratic government of Allende's predecessor, Eduardo Frei Montalva (1964–1970), indicate that his administration also considered relying on Soviet aid, and importing their industrial technologies for the realization of various infrastructural projects across the country. Based on a "Soviet loan that would allow Chile to promote its economic development through the acquisition of indispensable capital goods,"[18] this proposed cooperation was, however, blocked by new restrictions imposed by the Soviet authorities.

In political terms, the change in Soviet policy in regard to Chile stemmed largely from its pessimism (ultimately proved correct) concerning the viability of Allende's democratic socialist project, but commercial restrictions also qualified the nature of any aid provided by the USSR. At the 24th Congress of the Communist Party of the Soviet Union in 1971, delegate Aléxei Kosiguin explained to Chilean representatives that any cooperation was to be arranged within the context of mutual economic progress, framing the new policy of tied loans (for food and technical exchanges) under conditions identical to those of any agreement between capitalist market economies. The supposed qualitative differences between socialism and capitalism—as perceived by the Chilean left-wing—thus vanished.[19] As a consequence of these political and commercial strictures, Soviet policy regarding the export of its industrial technology also changed, favoring exchange rather than simply aid. Given these conditions, "Moscow refused to support heavy industrial projects in Chile, advising instead on the consolidation of light industries."[20] These light industries (of which KPD was an integral part) promoted a global socialist project while at the same time offering technologies not so advanced as to provide any succeeding government—which might have a different ideological agenda—an undeserved technological advantage.

During Allende's brief period in office, Chile had similar exchange relations with the UK. According to Alberto Arenas, then head of Chile's Sub-Department of Industrialized Housing, in 1972 British representatives approached him to offer a new system of industrial housing for developing countries called BRECAST. Two young Chilean engineers were immediately dispatched to England to learn about the system, since for Arenas, "it was a better system than KPD because it was more flexible and adaptable," and also it "didn't have the grandiose political associations of KPD."[21] The main difference between the two systems was that BRECAST—provided with a battery casting system—could produce precast concrete panels without the need to set up expensive factories.[22] With discussions arranging the exchange successfully concluded, at the beginning of September 1973 a BRECAST system was packed and ready to embark for Chile from a dock in Liverpool, only for Pinochet's coup d'état on September 11 to force the immediate cancelation of the order. This was not, moreover, the only shipment canceled. According to General Nikolai Leonov, Vice Director of the KGB's Department for Latin America between 1968 and 1972, at the precise moment of the 1973 coup three Soviet merchant ships were also heading to Chile, carrying weapons and tanks. When Moscow got the news, Soviet authorities gave immediate orders for the ships to turn around.[23] This double cancelation, for orders both infrastructural and military, gives some sense of the tensions that had existed for Allende in defending his democratically elected Unidad Popular against US embargoes with scarce economic support from the USSR—tensions that ultimately only allowed for one exchange venture to become a reality: KPD. The system sent to Chile was, in fact, the Cuban adaptation of the previous decade.

7.06 (left) Production
process in Cuban I-464 plant

7.07 (below) Plan layouts of
Chilean KPD housing blocks

First installed in November 1972, Chile's KPD plant ultimately went on to produce approximately 153 housing blocks installed in the cities of Quilpué, Valparaíso, Viña del Mar, and Santiago, and generated the equivalent of roughly 1,600 apartments a year. Each block typically contained sixteen apartments—six three-bedroom (of approximately 905 sq. ft.) and ten two-bedroom (720 sq. ft.). To date, no other industrialized building system in Chile has bettered either this annual production or the amount of internal space afforded to each flat. The factory employed 300 people working in two twelve-hour shifts to fabricate the panels, and another 400 were employed across various building sites, assembling the panels.

The plant's ideal radius for delivery was thirty miles; beyond this distance, delivery became uneconomic. Although only one KPD factory was ever in operation, the idea was mooted to relocate it when the need for housing stock in a certain region had been fulfilled, or even to introduce a series of factories throughout the country, with the plants to be converted into community centers once the necessary housing had been generated.

The Chilean government also gave broad support to developing and disseminating news stories, photographs, and films, celebrating the arrival of the KPD system and its industrialization of social housing. Alberto Arenas, a key figure in generating this propaganda, wrote a documentary on KPD that was to be shown on national television and in cinemas across the country. While all physical traces of the film are now lost, Arenas recalled that it discussed the housing shortage, and argued that "the city of the future" would require abandoning sprawl in favor of "more dynamic and dense projects":

> The idea was then to educate the masses by saying that the reality of living in a city would force them to forget having cats, dogs, chickens, and pigs, because society would demand a new way of behaving, and that we had to abandon customs that wouldn't be possible in dense high-rise constructions. We would then say that there were techniques now that could make this happen, and that the community, organized in teams, could develop a sort of cooperative property model where the individual inhabitant could begin to contribute to the making of the city by creating community spaces of co-ownership.[24]

FACHADA POSTERIOR BLOQUE 1

FACHADA INGRESOS BLOQUE 1

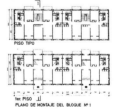

PISO TIPO

1er PISO
PLANO DE MONTAJE DEL BLOQUE Nº 1

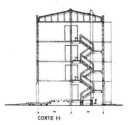

CORTE H

7.08 Production and
assembly process in Cuban I-
464 and Chilean KPD plants

The documentary also sought to alleviate fears of the new technology, emphasizing that it represented an important collaborative "unifying effort" between
Chile and the Soviet Union, and that it would allow higher buildings with "healthy
apartments, clean and affordable."[25]

A predictable consequence of this impassioned socialist—and therefore highly
politicized—message was that even before the 1973 coup, the KPD plant came to
be seen by opposing political factions as a malevolent Soviet enclave. According
to Patricio Núñez, architect and Sub-Director of the KPD plant, at a critical point
in early 1973 workers even began organizing committees to defend the factory

against possible attack, increasing surveillance day and night and running self-defense classes in the factory's forecourt. Increasingly, myths about the factory started to circulate, foremost among them the idea that the Russians were using the plant to camouflage missiles by disguising them as cranes, reviving fears of the Cuban Missile Crisis. Orlando Díaz, Vice-President of KPD's trade union remembers that a local newspaper:

> published a story that the KPD plant was not only a munitions factory but a factory that specialized in heavy weaponry. In 1973 they printed a picture of a missile that they said was being constructed at the plant. Anyone looking at the photograph would have clearly seen that it was actually a crane, but the guy who took the picture did so when the crane was at a 45-degree angle, as if it was ready to launch.[26]

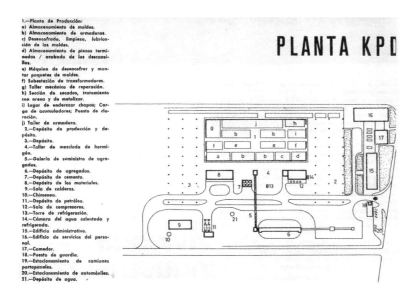

7.09 General layout of the KPD plant in El Belloto, Chile

7.10 Soviet cranes at the KPD plant misconstrued as missiles, November 15, 1973

Faced with this increasing barrage of criticism, the government upped its own propaganda efforts. "From the shovel to the button" became a slogan used to highlight the technological advances offered by the system. This idealized vision of automation was embodied by the KPD panel: as the key protagonist in a well-staged performance, this component was simply following a pattern of sequences dictated by a series of automated processes, in terms of both its production in the factory and its final assembly on the construction site.

Interestingly, the primary actors in this performance were not men but women, a role reversal recommended by the Soviets. As Arenas explained:

> The Russians asked us why we were hiring men to operate the cranes. They said that this is a very delicate job, and that women are much more meticulous at this kind of work, and if not already expert, they can be trained quickly. We immediately picked up on this idea and the women, in fact, did work very well . . . without the physical effort required to wield a shovel, a hammer or a saw, women could rely instead on their excellent hand-to-eye coordination.[27]

7.11 Journalistic report about KPD's female crane operators in Chile

7.12 The commemorative
KPD panel signed by
Allende as installed at the
factory entrance, 1972

7.13 The commemorative
KPD panel as modified
under Pinochet's
dictatorship, 1974

7.14 Photograph of KPD
worker, 1973–1974

This unexpected—and in Chile, quite unprecedented—subversion of construction site practice drew the attention of a number of illustrated newspaper reports. In one, María Elena Pivet observed:

> I am dressmaker by profession but I never liked it. Imagine, four kids to support. Now, though, they cannot believe that I work operating a crane . . . and they are very happy with what I do. Besides, I feel useful, working on something so important: building houses for Chileans.[28]

Another female crane-driver said, "we have all the men in the factory at our feet. We never feel more important than when we are sitting above them in the crane."[29] The integration of this female workforce was offered up as explicit evidence of the arrival of socialist gender equality. Like most of the workers in the factory, these women belonged to the Communist Party; the Party's

7.15 Photographs of KPD
workers, 1973–1974

7.16 The commemorative
KPD panel, photograph
taken on June 3, 2009.
Allende's signature still
remains underneath its
surface coating

ideological position perhaps explains the acceptance of women in this tradition-
ally male domain. Nonetheless, although Manuel Ramírez commented that it
was "all about respect, respect toward the female worker," María Elena Pivet
remembered that "sometimes it was necessary to beat certain guys up a little."[30]

As 1972 drew to a close, the limitations of Soviet economic support for Chile were
becoming increasingly apparent. As practically the only project to be realized
through the Soviet-Chilean alliance, the KPD plant clearly had symbolic power.
Only a day before traveling to Moscow in search of additional financial aid, Allende
had inaugurated the KPD factory with the sweep of his signature in the fresh
concrete of a new panel; this panel later became a commemorative object placed
proudly in front of the factory's main entrance. The inauguration was widely
publicized, and was important for the Soviets because it offered tangible proof
of the aid they had always promised. Allende took this achievement with him
to Moscow as evidence of the viability and success of Soviet investment in
Chile, and as representing the start of what he hoped would be further economic
and infrastructural support. However, he would return empty-handed, a disappointment that would quickly be overshadowed during the ensuing months by
the violent polarization of the country's political factions that culminated in
General Augusto Pinochet's US-backed coup d'état and in Allende's suicide on
September 11, 1973.

In the immediate wake of the coup, the KPD factory was taken over by the navy.
Most of its Soviet and Chilean executives were expelled from the country. Patricio
Núñez managed to keep his job for almost a year due to his technical and non-
political profile. Today, many ex-KPD workers believe that the large photographic
archive documenting the plant was confiscated and subsequently deployed to
identify perceived enemies of the ruling junta. Some KPD staff and workers were
detained, and in some cases tortured and murdered: in total, four were never
seen again.[31] As Susan Sontag has suggested, photography is indeed "an inventory
of mortality," stating "the innocence, the vulnerability of lives heading toward
their own destruction."[32]

Initially, navy officials searched in vain for the arsenal of weaponry supposedly hidden at the KPD plant. Alberto Arenas recalled that this search:

> resulted in the military tapping on panels and concrete flooring. When they thought they heard a weird or hollow sound they would break the slabs up with mechanical hammers to look underneath. At a certain moment Admiral Vargas came in and said, "This is crazy," and then went to talk to the person in charge. . . . This official said to me, "Look Alberto, . . . personally I think this technology is of the highest level and we should do everything we can to preserve it."[33]

Within days of the coup, Patricio Núñez met with Admiral Vargas, the new head of the facility who expressed his plan to reopen the factory using technical staff from the navy. Núñez recalled telling Vargas that this would be impossible:

> "If you want KPD to work again you need the people who know how to operate the machines; expert workers trained by the Russians. These people need to be brought back." . . . The day the workers returned to the KPD they were all called to stand in a line across the forecourt. I remember my feeling of shock when Vargas said, "I've brought you all back because according to my information you are indispensable. But you are now under my command, and it is my responsibility that you won't escape. If you do, not only you but your families will suffer the fatal consequences of your actions."[34]

By keeping the factory running, Pinochet's dictatorship also maintained the industrial paradigm pursued by the previous government. It seems as if the military junta valued the qualities of the housing system as much as the socialists. But in order to symbolically forgive (and erase) the "original sin" of the factory as expressed in its commemorative first panel, a process of ideological resignification was carried out: Allende's signature was covered up, and a religious altarpiece of the Virgin and Child was added, framed by two colonial-style lamp fixtures, recalling the iconography of a traditional Chilean country house altar. This rendering-over of a once-socialist emblem was supervised by Admiral Vargas himself. However, factory workers, displaying a certain awareness of history, first added a protective layer of oil over the concrete so as not to destroy Allende's signature.

With the plant's reopening, the initial, wide-eyed amazement at Soviet technology, and the introduction of women into the factory, turned into something quite normal, part of the daily routine. This is documented in the images taken by KPD's official photographer, Nolberto Salinas (remarkably, a man who survived Allende's overthrow to occupy the same role under Pinochet). In the images taken after 1973, the new technology is no longer the protagonist but appears as a secondary motif, as the backdrop. The photographic record, now in full color, is no longer used to present collective, institutional, and ideological stories, but instead constructs personal and seemingly intimate narratives which were sold to the workers as souvenirs. No longer recording and promoting a factory, the photograph becomes a rather intimate, depoliticized agent in an emerging family album.

The Chilean Construction Council, working closely with the authoritarian regime and empowered by its resistance to centralized building contracts as embodied by KPD, mobilized itself to secure the elimination of the KPD plant. In 1979, the factory was auctioned off, and then dismantled in the early 1980s. Today, the site belongs to a private pharmaceutical company, which has replaced the first KPD panel located at the factory's entrance with its own corporate logo. Only two concrete supports remain in the place where the first panel proudly stood in 1972—a panel that has only recently been found and saved from destruction by a former KPD worker.[35]

Notes

1 This chapter is part of an ongoing research project titled "Social and Industrial Culture of Soviet Large Concrete Panel Factories in Chile and Cuba: I-464 and the Politics of Prefabrication, 1958–1978" which has been sponsored by the the Vicerrectoría Adjunta de Investigación y Doctorado de la Pontificia Universidad Católica de Chile (2007), the RIBA (2008), the Getty Research Institute (2010), and the Canadian Centre for Architecture (2011). A version of this chapter was published in *AA Files* 59 (November 2009): 30–41.
2 Salvador Allende, interview with Régis Debray, "Allende habla con Debray," *Revista Punto Final* 126 (March 16, 1971): 57.
3 See Isabel Turrent, "El contexto internacional del experimento chileno, 30 años después," in *Frágiles suturas: Chile a treinta años del Gobierno de Salvador Allende*, ed. Francisco Zapata (Santiago de Chile: El Colegio de México–Fondo de Cultura Económica, 2006), 68–69; Isabel Turrent, *La Unión Soviética en América Latina: El caso de la Unidad Popular Chilena, 1970–1973* (Mexico City: El Colegio de México, 1984), 94; and Olga Uliánova, "La Unidad Popular y el Golpe Militar en Chile: Percepciones y análisis soviéticos," in *Estudios Públicos* 79 (Winter 2000): 100.
4 The first housing constructed using this method was part of a postwar reconstruction program in Le Havre, beginning in 1949. Docomomo-France, "Le Havre: Ilôt N17, système Camus, Fiche DOCOMOMO—English version," http://www2.archi.fr/DOCO MOMO-FR/fiche-havre-n17-va.htm.
5 Blair A. Ruble, "From Khrushcheby to Korobki," in *Russian Housing in the Modern Age: Design and Social History*, ed. William Craft Blumfield and Blair A. Ruble (Cambridge, UK: Cambridge University Press, 1993), 235.
6 Alexander D'Hooghe, "Siberia as Analogous Territory: Soviet Planning and the Development of Science Towns," *AA Files* 51 (2004): 14.
7 Julio García Luis, "La furia de las aguas," in *Revolución Cubana. 45 grandes momentos*, ed. Julio García Luis (New York: Ocean Press, 2005), 145.
8 Fidel Castro, "Una revolución es una fuerza más poderosa que la naturaleza," *Revolución*, October 23, 1963, reprinted in *Revolución Cubana*, 148–149.
9 Vitervo O'Reilly, interview with Hugo Palmarola, March 2011.
10 Fidel Castro, "Discurso del Dr. Fidel Castro en la Clausura del VII Congreso UIA," *Arquitectura Cuba* 331 (January–March 1964): 51.
11 O'Reilly, interview with Hugo Palmarola.
12 See Edmundo Azze, "Plan Director del Distrito José Martí (San Pedrito)," *Arquitectura/ Cuba* 337 (1967): 54–61; and Maspons González del Real et al, *Prefabricación* (La Habana: ISPJAE, 1987), 338–390.
13 Edmundo Azze, interview with Hugo Palmarola, March 2011.
14 Victor Rodríguez, interview with Hugo Palmarola, March 2011.
15 Ibid.
16 This poster appears in a series of slides taken of the Large Soviet Panel factory in Santiago de Cuba, and held by Patricio Núñez in his personal archive in Santiago de Chile.
17 Turrent, "El contexto internacional," 66.
18 *El Siglo*, June 1, 1971.
19 Joaquín Fernandois, *Chile y el mundo, 1970–1973: La política exterior del Gobierno de la Unidad Popular y el sistema internacional* (Santiago de Chile: Ediciones Universidad Católica de Chile, 1985), 359, 370.
20 Turrent, *La Unión Soviética*, 93.
21 Alberto Arenas, interview with Hugo Palmarola, April 2007.
22 Roger J. Lewis, "Mass Housing for the Third World," *The Architect* (June 1973): 36.
23 Nikolai Leonov et al., "El General Nikolai Leonov en el CEP," *Estudios Públicos* 73 (Summer 1999).

24 Arenas, interview with Hugo Palmarola. Although the script was completed, Arenas was unaware if the film was ever made or shown. There is, however, photographic evidence of a cameraman filming within the factory.

25 Ibid.

26 Orlando Díaz, interview with Hugo Palmarola, April 2007.

27 Arenas, interview with Hugo Palmarola.

28 María Elena Pivet, quoted in Jimena Castillo, "Mujeres al Volante de una Grúa: Reportaje a diez mujeres que decidieron reemplazar sus quehaceres domésticos por la mecánica, los atornilladores y las grúas," *Revista Paloma* 11 (1972): 34.

29 Irma Peralta, quoted in Castillo, "Mujeres al Volante."

30 Manuel Ramírez, interview with Hugo Palmarola, May 2007; María Elena Pivet, interview with Hugo Palmarola, May 2007.

31 *Detained and Missing: A Working Document* (Santiago de Chile: Archbishopric of Santiago, Vicaría de la Solidaridad, 1994). These four cases are mentioned in: *Tomo II: Casos Región Metropolitana*, Nelsa Zulema Gadea Galán (in 1973); *Tomo VII: Casos V Región*, Luís Gerardo Otárola Valdés, Sergio Jorge Hidalgo Orrego, and Hernán Leopoldo Quezada Moncada (all in 1977).

32 Susan Sontag, *On Photography* (London: Penguin, 1979), 70.

33 Arenas, interview with Hugo Palmarola.

34 Patricio Núñez, interview with Hugo Palmarola, April 2007.

35 In 2006, Servando Mora Cofré happened to be walking near the site of the former KPD factory at the very moment that the panel was being removed to make way for the new company's logo. The new owners allowed him to keep the panel, but only gave him until that evening to remove it. "At that time I was Secretary of the Agrupación de Exonerados Políticos de la ex-KPD, and so I phoned some fellow association members. . . . Together we arranged for a crane truck to remove the panel. . . . At the moment we are planning to construct a memorial. We have all the elements for this monument in place—an old KPD crane, a three-meter high statue of a worker made out of scrap metal and donated by the council, a new site in El Belloto and, of course, the panel itself, but we are still awaiting final approval and funds from the council." Servando Mora Cofré, "A Brief Personal Account of How the KPD Panel was Found," undated letter to Hugo Palmarola, received February 14, 2009.

8

ARGENTINA'S *CUESTIÓN CAPITAL*

Founding a Modern Nation[1]

Claudia Shmidt

With the ratification of the federal status of Buenos Aires on December 6, 1880, the so-called *cuestión capital* (the "capital" issue) of the Argentine Republic was resolved—leaving behind a toll of 2,500 dead and injured out of the 20,000 who had participated in the last and most violent chapter of the seventy-year dispute to define the seat of political authority. The battle for Buenos Aires was the end of a long confrontation between national and provincial governments, between civilians and the military, between political parties, between opposing figures, between different sectors of economic power. But it was also, without doubt, a fight for a *place*, for a *city*, that would bring together the material conditions necessary to house the federal government of an evolving modern nation.

It is important to remember that Buenos Aires was not always the capital of the Argentine state. Unlike other cities of the former Spanish Empire, it did not transition from viceregal to postcolonial national capital. The 1810 revolution brought separation from the Spanish Crown with the formation of the United Provinces of the Río de la Plata, and at once two conflicting models of the nation emerged: the *unitarios* (centralist) who wanted a single nation under a central government located in Buenos Aires, and the *federales* (federalists) who championed provincial autonomy. This schism provoked battles and strife that were only resolved in 1880. During these seven decades, the city of Buenos Aires—being the region's main commercial port—established itself as the capital of the richest province in Argentina, but it only alternately housed the national government, on a provisional basis, as a temporary guest. In practice, the asymmetrical relationship between this predominant port-city and the other provinces was always a cause of conflict.

Following independence from Spain in 1816, the violence intensified. In 1820, the Province of Buenos Aires declared its autonomy and began to organize its own administration, initiating important territorial, urban, and municipal transformations under the leadership of a distinguished group of politicians, businessmen, and intellectuals. However, the outbreak of war with the Brazilian Empire in 1825 over the territory of the Banda Oriental (now Uruguay) prompted the need to demonstrate national unity. Out of the same political group that had declared autonomy for Buenos Aires emerged the first President of Argentina, Bernardino Rivadavia; during his seventeen-month government (1826–1827), he approved the very first law for the demarcation of a national capital within a federal district. The urgency of the crisis laid bare the problem of not having a fixed federal capital, because a government lacking proper jurisdiction could not directly collect revenue and had to rely on subsidies from the provinces, thereby revealing its political and economic weakness. In the end, this law was not implemented. While the conflict with Brazil was resolved with the creation of an independent

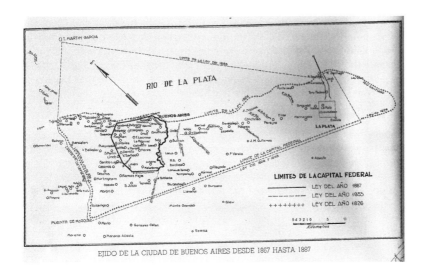

RIO DE LA PLATA

LIMITES DE LA CAPITAL FEDERAL
———————— LEY DEL AÑO 1887
— — — — — — LEY DEL AÑO 1853
++++++++ LEY DEL AÑO 1826

EJIDO DE LA CIUDAD DE BUENOS AIRES DESDE 1867 HASTA 1887

8.01 Political boundaries of Buenos Aires between 1867 and 1887 as represented by Martín Noel

Uruguay in 1830, the internal strife remained unabated, and Buenos Aires continued to maintain its dominance as a provincial capital through its control of customs and the port.

By mid-century, a general impasse permitted the ratification of the first Argentine Constitution (1853); nonetheless, the national capital remained undefined for three more decades. Long debates, violent confrontations, and political disputes ranged around the *cuestión capital*. During this period, the emerging institutions of the new republic operated from provisional capitals, housed in buildings constructed for other purposes, with transient locations and periodic relocations. Meanwhile debates and proposals concerning the characteristics, placement, and necessity for a capital for the new nation were underway.

In Spanish, the word "capital" has shifted in meaning, from an adjective qualifying the seven *pecados capitales* (Cardinal or Deadly Sins) in the Middle Ages, to a feminine noun (as in other Romance languages) denominating the main city and the seat of government. Following the United States Declaration of Independence and the French Revolution, the idea of a "capital" city acquired a particular connotation. It no longer referred only to the seat of kings or of governments, but became closely linked to the emergence of the "nation as novelty"—the construction of the modern state and all its various modes of political organization.[2] This would reframe of the tradition of a "capital-head" for a necessarily new political "body."

Through the example of Buenos Aires, I propose to revisit debates over the definition of the capital city from the viewpoint of theoretical political schemes that cross into the field of material representation, seen within the processes of nation-state formation in the mid-nineteenth century. In this context, Washington DC became an unavoidable model, even though many of the protagonists of the Argentine debates only knew it through the writings of Alexis de Tocqueville or other historical accounts. Washington DC functioned alternately as a referent to imitate or to reject, as synonymous with taming the wilderness, as the logical outcome of a civilized community, as "thinking head" in relation to other members of the national "body" such as New York or Boston. On the other hand, it was also viewed as an unworkable split for a country that aspired to combine economic and political power in a single federal center. Meanwhile, those looking toward

Paris, London, and other European capitals focused on the disjunction between their situation as royal capitals and the fascination they exercised as powerful and modern centers of progress. Examples from Latin America were discarded because they had not undergone the same conflicts over the *cuestión capital* as Argentina: Mexico City and Río de Janeiro were put aside due to the presence of monarchies, while Santiago de Chile and Montevideo appeared as the inheritance of the Spanish colonial past.

Initially, the lines of the debate showed commonalities with cities in other emerging nation-states that were experiencing similar processes of trans-forming themselves into capitals, such as Budapest, Prague, or Zagreb during the "invention" of *Mitteleuropa*, or even Nafplion (a short-lived experiment before Athens). As Ákos Moravánszky has pointed out, in contrast to the capitalist metropolis (such as Chicago or New York), the national capitals of Central Europe were more concerned with self-representation than with the efficiency of urban functions. In these cases, the coexistence of aristocratic and bourgeois institutions implied a collision of different visions.[3] Taking this context into account, the present study is guided by two questions: where should the city housing the federal government be located, and what form should it take? The question concerning the selection of a location would, of course, have parallels with later processes in India (New Dehli) or Australia (Canberra). In the majority of these examples, the debates spanned the range of options, from the extreme of the desert—framed as such due to its arid geography or to its absence of population—to the reformulation of existing cities; in the latter case, the question concerning the essential material features appropriate to a modern nation-state became even more acute.

Argirópolis: A Utopia

Following the international impact of the revolutions of 1848, a new stage began in the process of organizing the Argentine state under a republican system. One of the fundamental lines of this debate focused on how to include low-income populations: whether to control them underneath a strong centralizing power, or to incorporate them through education and the example of new European immigrants, who—it was hoped—would provide a "civilizing" influence as well as the knowledge and techniques of modern life. The preparation of the Argentine Constitution, which would strengthen the provinces as a group and would guarantee a government capable of centralizing public resources emerging from commerce as well as foreign credit, provided the decisive impetus.

In 1852, in the wake of the war over the Banda Oriental, Great Britain was finally forced to abandon its project of creating a new "mesopotamian" state—on the lands framed by the Paraná and Uruguay Rivers—which would have split up the Argentine provinces of Entre Ríos, Corrientes, and Misiones. The evidence that this political maneuver had failed was the success of Uruguayan independence, achieved thanks to the support of the powerful Brazilian Empire. Meanwhile, in 1850 Domingo F. Sarmiento (writing from his sojourn in Chile) had suggested the creation of Argirópolis as the capital of a proposed Federated States of the Río de la Plata,[4] to comprise the Argentine Confederation,[5] Paraguay, and Uruguay. A geopolitical rationale motivated this new union: Sarmiento maintained that landlocked Paraguay depended on the rivers linking it to the Atlantic in order to trade with Europe, and that Uruguay depended on the subsidies provided by France in exchange for its support for the continued French occupation of the island of Martín García—part of French efforts to maintain free navigation on the Paraná and Uruguay Rivers, in defiance of British efforts to control it. Sarmiento argued that they should share ownership of Martín García to site Argirópolis,

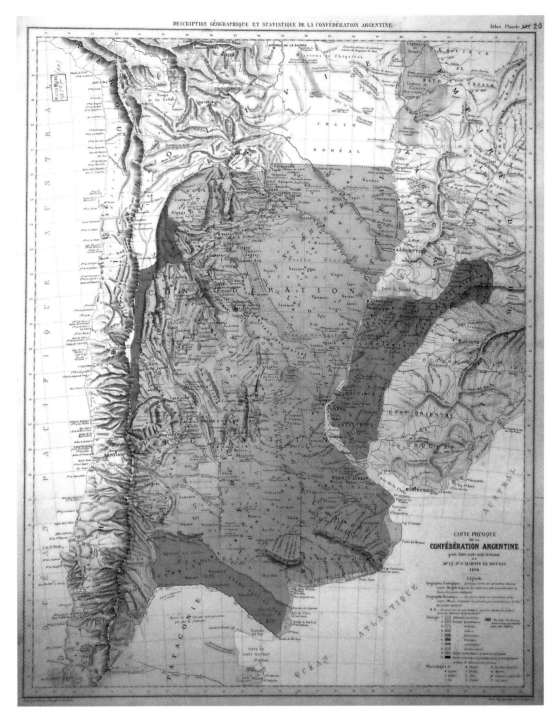

8.02 V. Martin de Moussy, map of the Argentine Confederation, 1869

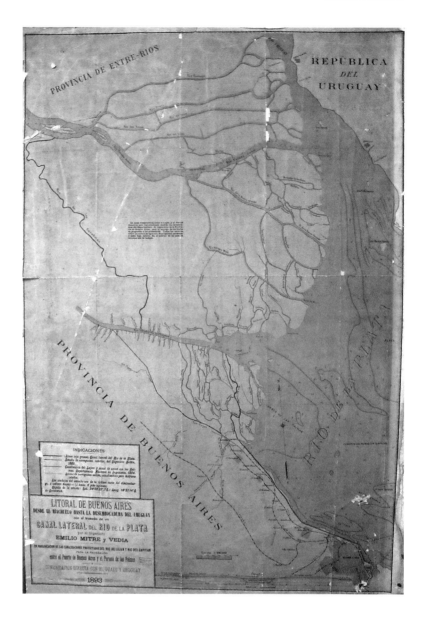

because this island was the key to commerce on the Paraná and Uruguay Rivers, and as a consequence, to the interests of the entire region. Furthermore, being an island granted it the character of a natural fortress.

Beyond the declaration of equality and neutrality, the vision of Argirópolis was charged with the tension of the *cuestión capital* in Argentina. Sarmiento argued that the United States of America was the model for modern republics, and that it had experienced similar difficulties in constituting itself as a union. Just like Buenos Aires (sited on the Río de la Plata), New York (sited on the Atlantic) was the wealthiest, most populous, and influential of the former English colonies, and for this reason the cities of Philadelphia, Baltimore, and Boston, among others, had tenaciously refused to allow an increase in its disproportionate influence by making it the seat of the federal authorities. With this comparison, Sarmiento reinforced the importance of the physical characteristics of Martín

García: not only would it *not* become New York (because in this schema, New York was Buenos Aires), but also being an island—in itself suggesting a symbol of independence—it would prove to be even better than Washington DC.

Argirópolis introduced a distinct variable by offering the possibility of conceiving of two separate "heads": one, a wealthy Buenos Aires that exercized "influence" over other cities, and the other, a different site to house the authorities, in the context of a regional organization—in this case the Federated States of the Río de la Plata, but if this were to happen under another political grouping, the capital would clearly belong to a separate geography. The function of the capital was to be the seat of a Congress that would defend the interests of its constituent states, and exercise strong fiscal, military–territorial, and above all commercial control over the whole. This plan did not propose a conventional capital, but a shared customs point for all the riverine settlements; it would be the "insuperable barrier" protecting a network of ports and cities located along the riverbanks, which would have to incorporate Córdoba through building a canal to the Tercero River. As an enclave at the crossing of the confluence of the river system, it would strengthen the connection with Europe, enabling a competitive advantage in relation to the rest of the continent.

A capital of this type could be "fabricated" rapidly. To prove his point, Sarmiento called attention to the extraordinary quantity of construction materials, houses and hotels of wood and galvanized iron, packed and ready for assembly, that had been imported into California over the last few years for exactly this purpose. Beyond this practical aspect, he was in large part inspired by impressions gathered on his recent travels. Stimulated by the heterogeneous ideas of the French thinker Victor Cousin, Sarmiento imagined a modern city that would manage to display a composition combining the best of Genoa ("its temples and buildings spilled over the steep slope of a mountain, not having but two streets in the entire city"), Venice ("founded on stakes in the heart of the lagoon"), or London ("would you like spacious, safe, convenient ports? Create 'docks' like those in London on the Thames").[6] Won over by Washington on his recent visit, Sarmiento situated the future Congress—just like the Capitol—on a natural promontory.[7] He imagined Argirópolis in picturesque mode: the mountainous nature of the island offered advantages over Washington since the varied terrain would break up the monotony of the landscape, and the elevated points would contribute to the fortifications. Argirópolis would emerge rich in construction materials: the rivers would bring to its ports wood from all over South America. It is important to note Sarmiento's conception of wood—and also of iron for certain kit homes—as "an element of durable construction" and of fast assembly (particularly if one considers that the Crystal Palace, the iron and glass pavilion built for the Great Exhibition in London in 1851, had not yet been built).

Clearly inspired by de Tocqueville's *Democracy in America*, Argirópolis offered the prospect of the triumph of progress with an even greater scope than a military victory, remodeling "the cursed desert" of this undeveloped terrain in Jeffersonian terms to promote a life of agrarian virtue.[8] The invention of Argirópolis even corresponds to the anticipatory visions of Brasília which had been articulated by José Bonifacio in 1821, in the context of the struggle for Brazilian independence[9]— an idea that re-emerged with the "dream" of Saint John Bosco in 1883.[10] Brasília in this sense shares with Buenos Aires decades of debate concerning the possibility of inventing a capital for a modern nation.

The Dilemma of the Capital City

The first Argentine Constitution (1853) stated in its third article that the national government of the Argentine Confederation would reside in Buenos Aires. It claimed that the State of Buenos Aires—which at that time did not even belong to the Confederation[11]—would cede its capital in order to be integrated into a federal system, a proposal that was immediately rejected by Buenos Aires. The outcome was an amendment to the Constitution that would postpone a definitive solution to the problem for three more decades: the city of Paraná was designated "provisional" capital with its province of Entre Ríos as federal district.[12]

Over these decades it is possible to identify two camps concerning the relationship of Buenos Aires to a national state: the "integrationists" and the "separatists." Within each camp there were also significant differences; the positions staked were not always clear, nor were the protagonists entirely coherent, since the complexity of the subject and of the interests involved obliged many politicians and thinkers to change their positions throughout the conflict. Among the "integrationists" there were two tendencies: those in favor of the city being the federal capital but dependent on its provinces, and those who maintained that it should assume a centralizing leadership role, exploiting its position of dominance. On the other hand, the "separatists" supported autonomy, defending the State of Buenos Aires—as much as the city—from those who would federalize it. More moderate positions suggested giving the city itself federal status while converting

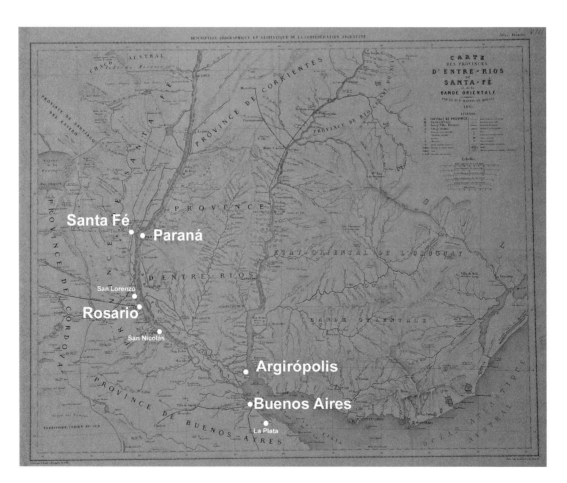

8.04 Diagram of the sites of possible capital cities between 1850 and 1887

the rest of the State of Buenos Aires into an additional province—a solution that, in effect, would be arrived at much later.

It can be said that Sarmiento and Juan Bautista Alberdi (author of the 1853 Constitution) marked two essential points on the changing and contradictory path of the debate over the capital in the second half of the nineteenth century. Both were convinced of the unsuitability of siting the capital in Buenos Aires, and both changed their opinion as political events unfolded, and also as its material, urban, and territorial transformations became evident. The idea of the need for a "capital" as the keystone of political organization belonged to a line of thought that had begun with the reflections of Machiavelli and Montesquieu, which became known through readings of their work by figures such as de Tocqueville or the Italian jurist Pellegrino Rossi, who were important points of reference for Argentine intellectuals. Rossi argued for the importance of a capital as the consummation of the centrality necessary for the government of a modern state.[13] His ideas were influential early on in the Unification of Italy—then unfolding— which was closely followed in Argentina, not only due to the similarity of the political situation and because it epitomized the longing for national unity as a form of progress and modernization, but also, fundamentally, due to the large presence of Italian immigrants who had fled poverty in Europe.[14] The counter-arguments, based on de Tocqueville, claimed that the concentration of power in a capital city would bring about new forms of despotism, maintaining outdated social structures characterized by hierarchical relationships of subordination.

Meanwhile, Buenos Aires itself became the apex of two conflicting political visions concerning the likely effects of centralization: one positive, that viewed it as stimulating and necessary, and the other negative, that saw it as an irrevocable path toward corruption and the loss of equitable control over power. In this climate, an unsettling question emerged: was it necessary *to fix* a capital? Some politicians asked why there should be a "permanent residence" for the authorities, and even less, a capital. Since the country had other urgent issues to address, why not resolve the issue with an interim site? Partisans who believed that it should be in Buenos Aires replied that the nation should unite, and did not have the resources to establish a new city: Buenos Aires was naturally, geographically, the "head" of the state.

While economic difficulties were afflicting the Confederation, the State of Buenos Aires maintained its autonomy, grew in strength, and modernized. The French-born engineer Carlos Enrique Pellegrini (whose son Carlos became President, 1890–1892), proposed in his journal *Revista del Plata* a model of progress inspired by the United States, promoting industrialization in order to supplement the river transportation system with one that included the entire region. It is notable that Pellegrini—who was a pioneer in the design of early iron structures, such as the first Teatro Colón in Buenos Aires—was a vehement opponent of the railway. He found that the effort to maintain the rectilinear and horizontal layout of the rails was irrational and contrary to nature, obliging the construction of dark tunnels that seemed to lead "toward death," or the elevation of banks or bridges over low-lying terrain: this was "the barbaric side of the current rails."[15]

In any case, in 1857 the first locomotive left Buenos Aires, heading west. Despite Pellegrini's pessimism, laying out the railway was key for future debates concerning the capital. Straight away, situating a capital on "the other side of the Paraná River" was rejected: its extreme width and the force of its current made the construction of a bridge too difficult. Meanwhile, the Confederation was still beset by political pressures and economic strains; in 1858, Entre Ríos ceased being the federal district, and the city of Paraná, as provisional capital, could barely continue this function for much longer.

8.05 (above) Carlos E.
Pellegrini, old Colón
Theater, 1857

8.06 (below) Buenos Aires,
Estación del Parque train
station, ca. 1878

In 1859, the State of Buenos Aires and the Confederation merged into the
Argentine Republic via the Pact of San José de Flores, and in 1860 they reformed
the Argentine Constitution. The third article was immediately repealed, opening
up a whole spectrum of possibilities for the capital. Instead of provisional locations
in provincial capitals, a permanent home for the national capital would be
established in an existing city or town, with the national executive given five
years to prepare the necessary buildings. In the first round, San Fernando and
San Nicolás de los Arroyos, north of Buenos Aires province, were proposed, but
neither was accepted. Eventually, a new solution was approved, even though it
was still provisional: the land housing the future capital should be ceded to the
nation by the affected provincial government.

This indeterminacy reoriented the debate. Some continued to support the idea of maintaining a provisional status for an itinerant capital: the city would be changed every five years, resulting in less corruption and greater equity. In 1862, the city of Buenos Aires was declared provisional capital for a period of five years, with the national government to provide whatever it needed, while public property would remain in the hands of the province. Thus it took up residence as a guest, until the stress of the situation exploded in 1880. Meanwhile, President Bartolomé Mitre emphasized that it was necessary to fix the site of a capital, recalling that the United States had launched "into the world" a "new and original theory concerning the capital"—that is, the invention of a new one, due to a lack of agreement over the existing options. However, he argued, an itinerant capital—moving from province to province—was completely unacceptable, since the situation in Argentina was not comparable to Switzerland and its cantons.[16]

The uncertainty regarding the location of a fixed capital meant that no one would construct public buildings, for fear of investing money in buildings that would have to be donated to the nation. The only permanent building that was national in purpose constructed during these years was the Customs House in Buenos Aires. This project emerged from a competition which had drawn three proposals. The winning project was literally situated on the water—on the Río de la Plata, to be precise—and it was known that the shores would always remain in the public domain, neither private not belonging to the provincial government. The building was a pioneer of its type in South America. Designed and built in 1854 by an English engineer, Edward Taylor—a follower of Jeremy Bentham's principles of functional organization—the building was laid out on a semi-circular plan and joined together various levels of warehousing orientated toward the river, with a pier for off-loading and for passengers. On the side facing the city was a lighthouse-tower, which for decades marked the urban landscape of the waterfront.[17]

8.07 Buenos Aires, view of the Customs House, ca. 1889

The Desert, the River, and the Railway: Sites for the Capital

In 1867, the nation's five-year term of residence in Buenos Aires expired. A year before, discussions on the old topic were renewed, but gaps had appeared between the world of speculation and the real transformations that the national state as a whole was driving, in the midst of ongoing political tensions. Although one of the central reasons for the selection of this site for the capital was directly connected to conflict over the control and distribution of revenues from the river ports linked to Europe, from now on increasingly the possibilities opened up by the railway—especially the greater supply of available land—would be incorporated into the debate. Progress in organizing a national, political front, based on the need to integrate the provinces, was consolidating itself, and by the time of Bartolomé Mitre's presidency (1862–1868), only two provinces continued to resist: Buenos Aires and Entre Ríos. The bid for localization became even more tense in view of this political map.

The 1866 proposal to establish the permanent capital in Villa Fraile Muerto in the Province of Córdoba initiated a cycle of bids from various cities in the interior. The alternatives oscillated between Córdoba, Rosario, Santa Fe, different points on the shores of the Paraná River—or even a point anywhere on a railway line. In any event, no alternative succeeded. On his return from the war with Paraguay, Mitre vetoed a law designating Rosario as capital. Sarmiento did the same a couple of years later during his presidency;[18] he also vetoed the project for a new city, Rivadavia, in Córdoba. Meanwhile the State of Buenos Aires maintained its provincial authority and the seat of the national government in its capital, in a coexistence whose conflicted nature only continued to grow.

The other topic that continued to circulate was the issue of size, a preoccupation that now acquired a more precise treatment—not a generic reference to the "large city" in the terms of Pellegrino Rossi, but an attempt to think about a measurable magnitude. Size proved to be an overwhelmingly functional consideration in arguments on all sides, whether for or against the location of the capital in Buenos Aires, in accordance with the weight of its power, its situation as a working port, and the modernity of its urban life in terms of commerce, transport, culture, and population. On one side, defenders of Buenos Aires such as José Evaristo Uriburu—then Minister of Justice and Public Education in Mitre's government—maintained that size mattered little, providing that its institutions operated freely: "No city is more free than London and yet it must be the largest in the world."[19] Those who rejected Buenos Aires argued that it would only contribute to the consolidation of a center to the detriment of the provinces, generating an intense imbalance: large capitals entailed grave dangers for people's freedom and for the independence of the national authorities. The relationship between "large city" and "freedom" took on various nuances, but when it was applied to concrete proposals, the representations of its evils or virtues translated into factors such as the distance or proximity of populated centers, the connection with railway or river networks, "palaces" or austerity in public buildings—forms which would all mold the "character" of a capital for a new modern nation.

But if London was the largest city in the world, what was the ideal size? As always in the *cuestión capital*, the obligatory reference was Washington DC. In support of Rosario's bid, it was argued that the US Constitution, far from mentioning the possibility of turning an existing city into a capital, on the contrary only discussed the selection of a district and limited its extension to ten square miles. By contrast, the third article of the 1853 Argentine Constitution established the possibility of turning a city into a capital, and did not limit in any way the land that should be attached.

The issue of size was specific to the local debate: the association of the "small" with the "new" was not given the same inflection in the debates over the definition of a capital for the new nation in Italy, for example. There the discussion was directed by the selection of an existing city such as Rome, bearer of history and, above all, of the power of the Catholic Church, which became a point of tension in the context of constructing a nation.[20] By contrast, in Argentina petitions to locate the capital in new towns such as Villa Fraile Muerto (transforming the older settlement of San Jerónimo), or Rivadavia (a new city to be superimposed over the layout of Villa María), were associated with their "lack of history." In these examples, the "new" was not seen as a projection toward the future, but rather as a guarantee that the capital would not be influenced by the past, and even less by established political traditions. The defense of Rosario was based on this same view, since at that time it was a small city. Those who did not agree used the same argument in reverse: if it became the capital, in less than ten years Rosario would reach a population of 100,000 and would suffer from the same evils as Buenos Aires.

Then how could a site be defined that would guarantee that the capital remained at a controlled size, physically and demographically? *Where* should this "new town" be located? The dilemma was framed by the "desert" on one side, and by the rivers and the railways on the other. For prominent Argentine politicians and intellectuals, the desert signified the danger of the "Indians"—whom they were systematically fighting—and also the dejection of large unpopulated areas. For this reason, the foundation of the modern nation would depend on inmigration policy.[21] Juan Bautista Alberdi, whose slogan was "to govern is to populate," had earlier pronounced that "any national government was impossible with the capital of the republic in Buenos Aires,"[22] even though he later confessed to having been mistaken, admitting that the invention of Washington DC would not work in the Argentine case, since it promised penal capitals in gloomy and deserted places selected to house a government that was prevented from being born, or that was being crippled or destroyed.[23] In this context, a military argument considering location in terms of geopolitical strategy brought one more alternative to the debate—although one that was latent, rather than recurring. In fact, the greatest fear was of internal violence and civil war rather than foreign invasion.

In 1871, upon vetoing the creation of a capital in Villa María, President Sarmiento outlined the difficulties associated with a desert site and the impossibility of constructing a new city in an "uninhabited" place. Far from his overflowing enthusiasm for Argirópolis, he now saw that building a new capital would take years, and argued that relocating the most prominent figures of Argentine society to live in improvised encampments would weaken the government's morale. Moreover it ran the risk of depleting the means of civilizing the countryside, leaving it to the mercy of the *caudillos* who dominated it from "the backwardness of the rural areas." Instead, Sarmiento presented a thorough plan to study the creation of a new city, in which there also appeared for the first time a possible program, in great detail, for a capital of national scope, which constituted, without a doubt, a novelty in the long years of this debate.

Beyond the concrete issues of the overbidding in order to retain political and economic power in Buenos Aires, this was not simply a debate over the selection of an existing place or city. Even Buenos Aires itself would have to be transformed if it actually were to be the national capital—but transformed into what? Because it did not have the buildings necessary to house a national government, therefore the investment required to realize this program would be, in effect, the same as for any site.

It is important to note that in this climate, Buenos Aires continued to think of itself as a modern city, with a view to consolidating itself as a future capital. In 1869, with the national government housed as a guest and having no greater jurisdiction, Marcelino Lagos—a lawyer, soldier, and journalist, who was very active in municipal affairs—presented a proposal to redefine the urban space of Buenos Aires via a circular perimeter boulevard, and a geometric urban layout using diagonal streets, plazas, and avenues. Lagos's project returned to a tradition of imagining the city that emerged out of ideas concerning the regularization of the street grid and the renovation of the existing city, present in various proposals of the time.[24] Furthermore, while this materialization of the boundary responded to an urban planning practice then at its height—as in the *ensanche* (enlargement) of important European cities—it was also evidence of the debate over size. Lagos left the water outside the city: he did not include the Riachuelo River, circumscribed the city's highest area, and—notably—interrupted the device of the circular perimeter in its intersection with the shore of the Río de la Plata. In addition, the Law of Expropriations had recently been promulgated, bringing greater legal

8.08 Buenos Aires, intersection between Cangallo and Reconquista Streets, ca. 1890

clarity in the disposition of land for public use. This made possible the planning of four equidistant plazas, organized by diagonal avenues, on the assumption that it would subsequently be possible to obtain the necessary land.

The power that Buenos Aires had acquired by the end of the 1870s made it into a force that was ever more difficult to counteract. No capital could be "better housed than in Buenos Aires, a city of all humanity due to the number of immigrants arriving, two-thirds of whom stay in its bosom, so it can thus be called a cosmopolitan city," said the politician Onésimo Leguizamón, whose position was always to promote a modernist and contemporary aesthetic in debates over the capital as well as in subsequent decisions about its public buildings. On the other hand, defenders of creating a new capital positioned these same values as a negative: the narrow streets of Buenos Aires (as well as Rosario) would be insufficient for the circulation of a growing population whose density would in fact require demolitions in order to open up wider streets and construct public buildings. Meanwhile prosperous cities bringing in huge profits were rapidly being erected in the eastern provinces of Entre Ríos, Corrientes, and Misiones. In conclusion, at this point the arguments which Sarmiento employed to reject the construction of a new capital in the "desert" were becoming more subdued due not only to the experience of growth sustained by new immigrant settlements, but also to the vertiginous modernization of Buenos Aires. In the north of Argentina, this "desert" was being transformed with the expansion of agriculture and of new cities.[25] In the south, out of the military campaigns to subdue and annihilate indigenous groups, Julio A. Roca emerged triumphant to become the protagonist of the final violent episode of the *cuestión capital*, which would then give way to a period of peace and consolidation of the republic.

The Definitive Capital

The key problem of the *cuestión capital*—the selection of the site—was finally resolved, by violent means. The denouement of this dispute—which, in its last stretch had irrevocably polarized forces within the Province of Buenos Aires in the face of conciliatory efforts and alliances among groups from the interior— reached its final point with a series of urban battles waged across the territory of the municipality of Buenos Aires and various strategic locations in its sur-roundings during June 1880. In this hostile climate, the engineers of the province's Department of Topography directed the construction of trenches in the city streets. One of them, the German land-surveyor Carlos Glade, managed to sketch out the construction of defensive walls around the city. The area to be protected, with dimensions comparable to Lagos's project, appears schematically in a drawing marking—at a distance of approximately one kilometer apart—a series of eight bastions connected by a broken line.[26] A final gesture of desperation to retain a city which at that point still belonged to all.[27]

Finally, the fall of the governor of the Province of Buenos Aires, Carlos Tejedor, and the rise to power of Roca, precipitated the eventual agreement. The solution consisted of ceding to the nation the territory of the city of Buenos Aires, which was declared the Federal Capital, and granting federal funds to build a new capital—but this time, to house the provincial government. Immediately two processes of urban development began in parallel. On one hand, the provincial government expedited the formalities of deciding the placement and the characteristics of its new capital, La Plata. Located in direct relation to the port of Ensenada, its modern layout on a grid crossed with a system of diagonals materialized in record time, along with the construction of its most emblematic public buildings, in most cases designed through an international architectural competition.[28]

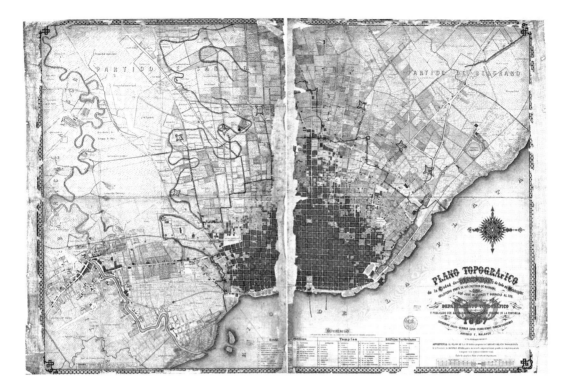

8.09 (above) Topographic
map of Buenos Aires upon
which fortifications were
sketched in 1880

8.10 (right) A 1888 map of
the final political boundaries
of the Federal Capital of the
Argentine Republic. The
darkened area corresponds
to the extent of the
municipality of Buenos Aires
before federalization in
1888

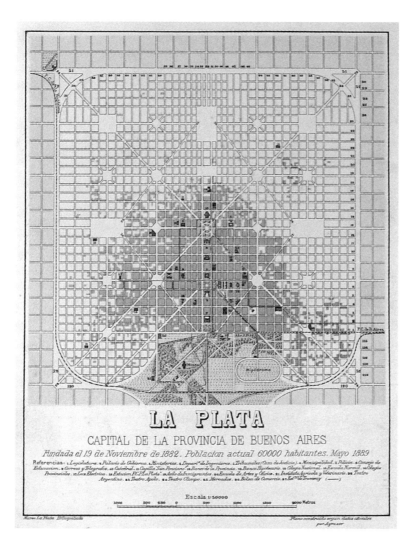

LA PLATA

CAPITAL DE LA PROVINCIA DE BUENOS AIRES

Fundada el 19 de Noviembre de 1882. Poblacion actual 60000 habitantes. Mayo 1889

8.11 Plan of the city of La Plata, founded in 1884 as the provincial capital of the State of Buenos Aires, showing the extent of its buildings

At the same time, the triumphant national government quickly undertook the transformations to adapt Buenos Aires to its new function as capital. The definitive capital brought to an end years of provisional solutions, not only in political terms: the installation of republican authorities required permanent monuments, as Roca clearly expressed at the beginning of his administration. These monuments should be for the entire nation, therefore the capital had to belong to the republic, and it is in this sense that (along with a sector for the elite) Roca fought for the considerable expansion of its territory.[29] Accordingly, a few days after the ratification of the city's federal status, a municipal official submitted a request to regularize its boundaries, which required the province to cede a few additional square meters of territory. What began as a simple ordering of the boundaries ended with a petition for enlargement that tripled the area of the federal capital and the layout of its definitive form. During this process, which took six more years, Buenos Aires organized the administration of the state, and undertook a number of important public works projects—developing the layout of the railway network, and constructing a controversial new port, where political interests overcame technical considerations in the selection process.[30]

8.12 Diagram of the railway system of the Argentine Republic superimposed on a map of Great Britain to show its scale, 1888

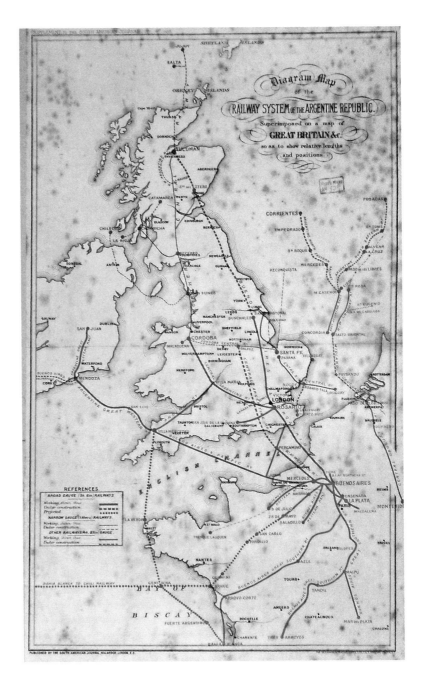

But while it had only just managed to settle on the site and propose designs for "public buildings worthy of the nation"—seats for the legislative, executive, and judicial powers—which would take several more decades to build, it did succeed in realizing one of the most notable undertakings of the era in Latin America: the construction of true educational "palaces"—building sixty-four of these new schools in just four years. On May 25, 1886—five months before completing his term—Roca inaugurated forty "palatial" public school buildings in a single day. With the 1881 creation of the National Education Council and the legal institution of free and compulsory public education, the Argentine state could now display in its brand-new capital a concept that was integral to the modern nation: an education system that proposed to create a homogeneous population out of native-born residents and the anticipated European immigrants, imparting training in the values of the new nation, the most advanced principles of hygiene, and a productive administrative system that would position the state in the context of the global economy.

On January 17, 1888, two engineers—Pablo Blot nominated by the national government, and Luis Silveyra by the provincial—drew up the definitive plan for the new federal district, with the layout of a "wide boulevard for circulation." The "great capital" in the "great city" of Buenos Aires became a reality, along with the

8.13 (left) View of Buenos Aires showing the old Congress Building, ca. 1880

8.14 (below) Francesco Tamburini, project for the National Parliament Building, 1885

8.15 Carlos Altgelt and the
Consejo Nacional de
Educación, Petronila
Rodríguez School, Buenos
Aires, 1882–1888

8.16 Carlos Morra and the Consejo Nacional de Educación, Onésimo Leguizamón School,
Buenos Aires, 1884–1886

consolidation of a federal republic with a centralized government. Its territorial expansion and the development of unprecedented public works constituted a phenomenon that superimposed its own standards, intersecting the processes of modernization in the context of the construction of the new nation-state.

Notes

Translated from the Spanish by Helen Gyger, with Patricio del Real.

1   This chapter forms part of my doctoral dissertation, to be published as: Claudia Shmidt, *Palacios sin reyes. Arquitectura pública para la capital permanente 1880–1890* (Rosario: Prohistoria Ediciones, forthcoming).

2   "The basic characteristic of the modern nation and everything connected with it is its modernity": thus Hobsbawm begins the chapter "The Nation as Novelty: From Revolution to Liberalism," presenting the historical development of the idea of the nation and its resignification after 1884 when it became connected to the idea of the state. Hobsbawm argues that previously "nation" had meant the group of inhabitants of a particular country, or—following its etymology from French *naissance* (birth, source)—the birthplace. Eric J. Hobsbawm, *Nations and Nationalism since 1780: Program, Myth, Reality* (Cambridge, UK: Cambridge University Press, 2004), 14–15.

3   Ákos Moravánszky, *Competing Visions: Aesthetic Invention and Social Imagination in Central European Architecture, 1867–1918* (Cambridge: MIT Press, 1998).

4   Domingo F. Sarmiento, *Argirópolis o la Capital de los Estados Confederados del Río de la Plata* (Santiago de Chile: Imprenta Julio Belín y Cía, 1850).

5   Initially called the United Provinces of the Río de la Plata, in the period 1835–1852 the Argentine Confederation comprised all the provinces and territories except for the Province of Buenos Aires.

6   Sarmiento, *Argirópolis*, 83.

7   Ibid., 90.

8   Natalio Botana, *La tradición republicana. Alberdi, Sarmiento y las ideas políticas de su tiempo* (1984; Buenos Aires: Sudamericana, 1997), 291–292.

9   José Bonifacio, "Lembranças e apontamentos de Governo Provisório para os Senhores Deputados da Provincia de Sao Paulo," in *José Bonifacio*, ed. Pedro Pereira da Silva Costa (São Paulo: Editora Três, 1974), 106.

10 Saint John Bosco, the Italian saint born in 1815, related a dream in which a land of well-being arose between the fifteenth and twentieth parallels, very close to a lake—a location consistent with Bonifacio's specifications.

11 Between 1852 and 1862 the Province of Buenos Aires changed its name to the State of Buenos Aires in order to emphasize its separation from the Argentine Confederation.

12 During the period that the Constituent Assembly was in session in 1853, eleven laws were introduced in connection to the *cuestión capital*. Arturo B. Carranza, *La Cuestión Capital de la República Argentina. 1826 a 1887. Antecedentes, Debates parlamentarios, Iniciativas, Proyectos y Leyes*, 4 vols. (Buenos Aires: J. Rosso, 1926–1932), 2:54.

13 Pellegrino Rossi was following Montesquieu, "Mes pensées," in *Oeuvres complètes*, 887, quoted by Botana, *La tradición republicana*, 151.

14 Fernando Devoto, *Historia de la Inmigración en la Argentina* (Buenos Aires: Sudamericana, 2003).

15 Carlos Enrique Pellegrini, *Revista del Plata* no. 5 (January 1854).

16 Carranza, *La Cuestión Capital*, 2:172.

17 Daniel Schávelzon, *Haciendo un mundo moderno: la arquitectura de Eduard Taylor (1801–1868)* (Buenos Aires: Ediciones Olmo, 2010).

18 Sarmiento was President of Argentina, 1868–1874.

19 Carranza, *La Cuestión Capital*, 4:298.

20 Alberto Caracciolo, *Roma capitale. Dal Risorgimento alla crisi dello stato liberale* (Rome: Edizione Rinascita, 1956).

21 Tulio Halperin Donghi, *Una nación para el desierto argentino* (Buenos Aires: Centro Editor de América Latina, 1992); Tulio Halperin Donghi, *Sarmiento: Author of a Nation* (Berkeley: University of California, 1994).

22 Juan Bautista Alberdi, *La República Argentina consolidada en 1880. Con la ciudad de Buenos Aires por capital* (Buenos Aires: Pablo Coni, 1881), 75.

23   Juan Bautista Alberdi, *Escritos Póstumos*, vol. 9 (1861), 161, quoted in Botana, *La tradición republicana*, 404.

24   The projects for Buenos Aires from this period have been analyzed extensively, in particular by Adrián Gorelik in the first chapter of *La grilla y el parque. Espacio público y cultura urbana en Buenos Aires, 1887–1930* (Buenos Aires: Universidad Nacional de Quilmes, 1998), 57–100.

25   Jorge Francisco Liernur, "La construcción del país urbano," in *Nueva Historia Argentina. El progreso, la modernización y sus límites (1880–1916)*, ed. Mirta Z. Lobato (Buenos Aires: Sudamericana, 2000), 5:179–208.

26   The scheme was no more than a sketch in carmine graphite, drawn on an old plan of the city (by Malaver, from 1867) but it indicates a knowledge of the city as it was in 1880.

27   Hilda Sabato gives a detailed history of the final stages of the conflict, which took place between 1879 and 1880. Hilda Sabato, *Buenos Aires en armas. La revolución de 1880* (Buenos Aires, Siglo XXI, 2008). See also Hilda Sabato, *The Many and the Few: Political Participation in Republican Buenos Aires* (Stanford: Stanford University Press, 2001).

28   Fabio Grementieri and Claudia Shmidt, *Alemania y Argentina. La cultura moderna de la construcción* (Buenos Aires: Ediciones Larivière, 2010), 193–251. The publication includes an English translation of the text.

29   Claudia Shmidt, "Palaces with no Kings: Architecture for Public Buildings in Buenos Aires, 1884–1906," in *Architectural Culture around 1900: Critical Reappraisal and Heritage Preservation*, ed. Fabio Grementieri, Jorge Francisco Liernur, and Claudia Shmidt (Buenos Aires: UNESCO World Heritage; Universidad Torcuato Di Tella, 2003), 110–117.

30   Between 1882 and 1887 there was an intense debate over two rival proposals for the port, devised by Argentine engineers representing different political and economic interests. The winning project was by Eduardo Madero, an ally of Roca, but it was not the best from a technical standpoint. It never functioned particularly well, due to its placement of the docks parallel to the shore—which reduced entry and exit to only two fixed points—instead of a perpendicular alignment, as engineer Luis A. Huergo had proposed, and which was eventually built in an adjacent area, called Puerto Nuevo, in 1912. See Jorge Francisco Liernur, ed., *Case: Puerto Madero Waterfront* (Munich; Harvard: Prestel, 2007).

9
MODERN FRONTIERS

Beyond Brasília, the Amazon[1]

Paulo Tavares

By the time Brasília was officially declared the political capital of Brazil in 1960, a massive process of territorial reorganization had started to unfold in the Amazon. Observed at territorial scale, the transversal axes that define the basic urban structure of Brasília appeared as an intersection of two major arteries that were being cut toward the northern and western borders of Brazil, one linking Brasília to the city of Porto Velho, near the border with Bolivia, and the other connecting the new capital with the city of Belém, the main port situated at the delta of the great Amazonas River. Roads were designed as lines of occupation as well as channels through which modernization would infiltrate the undomesticated interior. For the government, as well as for many geographers, engineers, and architects, it was necessary to reorient the spatial axis of coastal occupation inherited from the colonial period toward the central plateaus, and from that elevated point—Brasília—unleash a new cycle of development that would expand to the rest of the national territory.[2]

During the twentieth century, when Brazil moved away from a rural-based economy with increasing urbanization, territorial occupation and control acquired the sense of a manifest national destiny and civilizing mission. Whereas the natural wealth of the country's interior would provide the source of capital accumulation to fuel the industrialization of the economy, the "construction of

9.01 Ground zero of Brasília: two axes crossing at a right-angle define the basic urban structure at an early stage of construction, 1957

the territory" was transformed into a symbolic space through which the image of a modern nation, technically capable of mastering its geography and sovereign of its own resources was being created. Lucio Costa, urban planner of Brasília, described the rationale for the new capital clearly: "It was born out of the primary gesture of one who marks or takes possession of a place: two axes crossing at a right angle; the very sign of the cross."[3] Brasília's modern message is only fully grasped at the intersection between urban and territorial scales, precisely at the moment when its CIAM-informed design meets the reenactment of a colonial gesture of territorial occupation.

Only in the early 1970s, when Brazil was under military rule, did Brasília become the de facto center of political power and have its adjacent roads paved. By that time, an entire new map of the Amazon was being drawn. After the 1964 military coup, frontier expansion acquired the rationale of a centralized, state-led process of "territorial design" implemented through successive development programs. On the ground, these projects would take the form of an urban matrix of continental proportions, formed by a series of modern enclaves interlinked by communication and transport lines cutting throughout the entire Amazon basin. This planning scheme was elaborated by a multilayered discourse that combined geopolitical strategy, modernization theories, regional planning diagrams, and the vocabulary of modernist architecture and urbanism that had already been consolidated with Brasília. Similarly to the new capital, the "construction of the Amazon" played both an economic and a symbolic role, acting as the material reserve that drove the high levels of GDP-growth experienced under the military regime, as well as constituting a space through which a hegemonic national imaginary that helped to legitimize authoritarian and violent forms of state control was constructed.

This chapter examines several spatial probes of the process of modern colonization of the Amazon up to late 1980s, when the "nationalist-developmental" project that animated the expansion of the western frontier became fractured. By 1989, when environmental discourse had already infiltrated public sensibility and Brazil was running its first democratic elections in twenty-five years, the Amazon emerged as a multi-scaled arena of political conflict wherein ethical claims on behalf of humanity and nature were inextricably articulated with the contingency of local socio-ecological histories affected by the forces of global modernization. Throughout the Third World, spaces that supplied raw materials for the militarized industrialization experienced during the Cold War were converted into transnational lines of dispute around which local communities and an expanding international network of environmental advocacy and activism gathered in defense of dissident modes of managing natural resources. In the Amazon, at the fringes of the modern-colonial frontier, hegemonic articulations between national imaginaries, the state, and the territory were contested and reshaped at the moment when democracy was being defined anew.

Primitive Islands

In 1956, the architectural and arts magazine *Habitat* published a sequence of three photoessays about the Karajá, an indigenous group that lived about three hundred miles north from where Brasília was being built. Created by Italian-born architect Lina Bo in 1950, *Habitat* was one of the most important instruments through which modernism was debated and propagated in Brazil after the Second World War. Like other architecture magazines that emerged at this time, most importantly *Módulo* created by Oscar Niemeyer in 1955, *Habitat* not only circulated images of the European avant-gardes and the vibrant modernist production gestated in Brazil since the 1930s, but also brought to the urban elites of Rio de Janeiro and São Paulo the images of a territory that, while being constructed,

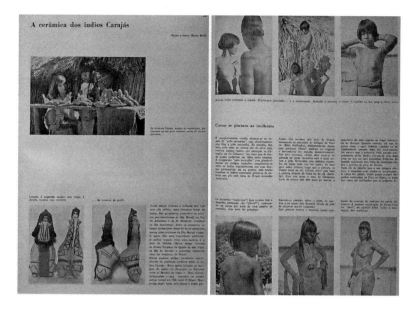

9.02 Modern primitives: the Karajá as they appeared in the photoessays by Mario Baldi in *Habitat*, 1956

was also being rediscovered. To be modern—these modernist media pedagogically communicated—was to be connected both with the language and ideals of modernism and with the native and popular forms of cultural expression.

The articulation between the vernacular, indigenous, and the modern that circulated through these publications is representative of an anthropological gaze that was not only a structural element of Brazilian modernism in general—not to say of modernism *tout court*—but also the main axis around which multiple notions of modernity and national identity were formulated and disputed. Since the first decades of the twentieth century, when nationalism emerged as a central force in the political scenario of Brazil and modernist ideas became increasingly present in the circles of urban elites, images of the native and local culture were appropriated by various intellectuals, artists, writers, and architects. The "discovery of the territory" was mobilized by aesthetic projects that occupied different positions on the ideological spectrum, serving both for critical perspectives toward the political and cultural structures inherited from the colonial period and for privileged positions that simultaneously redefined and reinforced them.[4] Arguably, all of them, from left to right, were directly or indirectly influenced by a similar imaginary resource, which was provided by the circulation of photographs, films, and artefacts collected and recorded during all sorts of exploratory incursions, scientific–military expeditions, religious missions, and photojournalistic reportages that penetrated toward the interior of the territory at that time. The Karajá photoessays are interesting, therefore, not so much because of what they exhibit about the Indians, nor solely because of the modern message they convey, but most importantly because of the historical circumstances that allowed their emergence and circulation, which although not immediately visible, reveal the image of a frontier-space that was in itself modern.

Mario Baldi, the Austrian photojournalist who authored the articles, was involved in photographic expeditions and collaborated with many other publications after migrating to Brazil in 1921. From the early 1930s, he joined official expeditions organized by the Service for the Protection of the Indians (SPI) and traveled on several occasions to the territory of the Karajá,[5] an indigenous group with whom contact was recorded as early as the seventeenth century with Jesuit missions.

9.03 (below) A modern
frontier: an airfield used by
the "pioneers" of the
Roncador-Xingu Expedition,
ca. 1950. After first contact
was made with indigenous
tribes, the second step was
to open up an airfield and
outpost for future
incursions. Many
settlements established in
this manner turned into
cities

9.04 (facing top) Pacifist
and expansionist: the
Kuikoro indians stand in
line to receive gifts from
the filmmaker Nilo Oliveira
Vellozo during a SPI mission
to the Culuene River basin,
1944. "Gift-giving" was a
widely used tactic in making
contact

9.05 (facing bottom) First
Mass: the inauguration of
Brasília was marked by a
Catholic Mass on May 3,
1957. Oscar Niemeyer
designed the temporary
chapel to accommodate
people from all over the
country, including
representatives of the
Karajá, who were brought
from the Ilha do Bananal by
the Brazilian Air Force

When the photo reportages were published in *Habitat*, the river margins inhabited by the Karajá were being prepared for yet another wave of spatial transformation brought by the construction of the highway linking Brasília to the city of Belém, in the delta of the Amazonas River, which was planned to cut through the middle of the territory they inhabited. Although Baldi's ethnographic-informed narrative portrays the Karajá as an almost isolated cultural entity and erases much of that historical–political context, traces of it emerge in a short passage of the text: "Before, they used to live dispersed in many villages. In 1936, I visited 27 of them. . . . Today, however, all of them or at least the great majority live on the Island of Bananal, where they have been concentrated by the SPI."[6]

The multiplicities of peoples which these expeditions encountered, all of them generically characterized as "Indians," were placed in an ambiguous relationship to the modern nation-state that was being territorially constructed. Insofar as they were considered as part of the nation and "potential citizens" who had to be integrated into society, their existence and presence functioned to legitimize civilizing missions toward the interior of the territory. On the other hand, the dispersed Indians constituted an obstacle for the expansion of the frontier itself, since they obviously offered resistance to the occupation of their lands. Created in 1910, the SPI—also known as SPILTN, which added to the acronym the phrase "and Localization of the National Worker"—eliminated this contradiction through a form of practice that was pacifist and expansionist at the same time. The agency advocated for a non-violent form of tutelage held by the state over the indigenous people, circumscribed inside "protection islands" where dispersed communities were resettled after being pacified. While operating as contact posts through which the Indians would be pedagogically integrated into the national society,

these "indigenous parks" also served as buffer-zones to preserve their culture and ensure their survival against armed ranchers, miners, land-grabbers, viruses, and other effects brought alongside the frontiers of modernization.[7]

The preservationist necessity of collecting, archiving, and diffusing the "cultural heritage" of indigenous and popular culture was the reverse symptom of the intensification of the expansionist policies promoted by the government of President Getúlio Vargas. After seizing dictatorial powers in 1937, Vargas launched a large-scale program of territorial occupation that included many efforts to penetrate into the interior. The "March to the West"—as it was officially known—actively promoted the migration of settlers, enlarged transport and communication networks, and supported the expansion of the exploratory–pacifying missions initiated with the SPI.[8] While modernization was directly identified with the occupation of the interior, the very image of the frontier acquired a symbolic significance of its own. Humanitarian in intent but governmental in fact, the politics of pacification and confinement directed toward indigenous communities allowed the domestication of their images, inscribing the primitive as part of the national-modern imaginary while isolating them from any form of political agency.

Modern Enclaves

Although much of the imaginary about the Amazon constituted a symbolic resource of primitiveness and the vernacular, the Amazon is a territory where modernity has historically found a fertile ground for experimentation. One of the most expressive and pioneering examples is Fordlandia-Belterra, the urbanized rubber plantations built by the Ford Motor Company in a remote area at the margins of the Tapajos River. Part of a geopolitical strategy of the US economic interests to circumvent the monopoly of the rubber trade from British-controlled colonies in Southeast Asia, Fordlandia (1927), and later Belterra (1934), like their twin sister Harbel built by Harvey Firestone in Liberia in 1926, were inscribed in a global geography that extended the rationality of the factory's assembly lines to frontier spaces of the Third World. The mechanized plantations were equipped with all the architectural facilities and institutions necessary to run a city, precisely arranged according to the spatial divisions and hierarchal schemes of capitalist industrial production.[9] And yet, despite Fordlandia's functional design, its modern infrastructures, and the ideology of progress and modernization it embodied, the plantation was absent from the images of the Amazon that later circulated in architectural magazines. The Anglo-American bungalow style of the city and its suburban atmosphere did not provide suitable references for the vocabulary promoted by modernist architects, but there was perhaps no other urban configuration before the construction of Brasília that expressed so precisely the project of occupying and modernizing the interior. The city-plantations combined frontier conquest, capital-intensive development, and modern rationalities of government into a single structure; at the same time, taming the wilderness of nature and projecting new values and behaviors onto what was otherwise considered a non-civilized territory/population.

9.06 (opposite top) Modern enclave: isolated by the forest, Vila Serra do Navio was designed by Oswaldo Bratke to be "an exemplary city," 1955–1960

9.07 (opposite below) Oswaldo Bratke, Civic Square at Vila Serra do Navio, 1956

From 1955 to 1960, concurrently with the construction of the new capital, another modern enclave was built from scratch near manganese-rich hills situated at the north of the Amazonas River delta. Designed by architect Oswaldo Bratke, a central figure of Brazilian modernist architecture, Vila Serra do Navio (VSN) was the more important of two urban nodes of a large mining complex that was implemented by US giant Bethlehem Steel in partnership with a Brazilian mining company named ICOMI and with the financial support of the World Bank. The extractive infrastructure comprised the installation of an urban compound adjacent to the new river port, a railway link with the mining site that was remotely located 125 miles deep in the jungle, and next to the mine, a city for 3,500 residents was constructed to house workers, technicians, and administrative personnel.

The urban structure of VSN reproduced the division of labor inscribed in the process of extracting manganese. The employees of the company were placed in different types of accommodation organized in two large isolated sectors: one for the workers and the other for "personnel of medium level" and "graduate

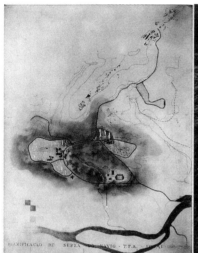

professionals." The workers' housing was further subdivided into two sections: clusters of twin-houses for families and a major communal housing block allotted for single male employees. At the center of the housing blocks, Bratke placed the school, and what he called the "civic square"—a larger architectural complex equipped with services, commercial, and leisure facilities. On the opposite side, separated from the workers by a long green park, a similar layout housed middle- and upper-rank personnel, who themselves were divided into two separate blocks distributed around a smaller center where the architect located the administrative complex, a hotel for visiting professionals, and collective housing for female workers, who were mostly employed in domestic service in that area of the town. "The localization of the housing for single workers," Bratke noted, "assumes a moral character in a place of monotonous routines."[10] To assure that the disciplinary mechanism of the city would properly function, when admitted to work at VSN, each worker received a manual that established some basic rules for the use of public spaces and urban amenities, including the maintenance of the houses. Whereas Bratke acknowledged that the urban scheme implied a certain segregation, the architect contended that it was the most functional, and that all the employees shared a common living standard provided by the modern design applied at both poles of the city. At once spatially divided but symbolically unified, modernist architecture would create homogeneity between the inhabitants and generate a sense of community throughout a hierarchically segmented urban fabric.

VSN was a product of the militarized economies that emerged out of the war. One of the main suppliers of the Allied forces, Bethlehem Steel expanded its productive basis following the enlargement of the US domestic economy and the increasing demands for steel-based products necessary to rebuild war-torn economies. By the mid-1950s, when industrial strongholds such as Germany and Japan had industrial areas under reconstruction, Bethlehem Steel's output was at its highest level and the company was expanding its global outreach toward new sources of raw materials.[11] Bratke visited many corporate resource-extraction complexes in Latin America before presenting his urban planning proposal, including a company town named El Pau that was built by the company in Venezuela five years earlier. Although these cities offered an adequate organiza- tional model, Bratke insisted that their spatial arrangements lacked the necessary civic character that was integral to urbanity.[12]

The architect also conducted a series of visits to the Amazon during which he collected photographic material, sketches, and notes about the living conditions in riverside communities of migrant workers and settlers. These settlements and the houses themselves were most probably another end-product of the war- effort: after Japanese troops took control of the rubber plantations in Southeast Asia, the Brazilian government, with the financial support of the United States, organized a massive immigration program in order to increase rubber extraction in the Amazon. Around fifty thousand men were displaced to the region, most of them impoverished peasants from drought-plagued areas of northeast Brazil. Throughout the rubber booms and depressions of the nineteenth and twentieth centuries, these migrants had developed a particular architectural typology alongside the margins of the Amazonian rivers that held a certain fascination for modernist architects. The first issue of *Habitat* included an extensive photoessay on these self-built houses entitled "Amazonas: The Folk Architect." *Módulo* also dedicated space in its first issue to a short essay about the region's vernacular that was signed by the engineer Joaquim Cardozo, Niemeyer's long-term collaborator. Similar to the "pile dwellings" of the Congo depicted in Bernard Rudofsky's *Architecture without Architects* ten years later, the architectural simplicity of these houses was equated with the sophisticated grammar of modernism. The

9.08 Anthropological gaze: the architectural vernacular of the Amazon as it appeared in *Habitat*, 1950; articles entitled "Building and Living" and "Amazonas: The Folk Architect" illustrated by images of indigenous structures

crude and minimal aspect of the dwellings, their long and open collective decks, their "architecture à *pilotis*" structure,[13] and the environmental responsiveness of their design were framed as references for the construction of a national-modern imaginary alongside the graphic geometry presented in images of indigenous artifacts. While Bratke's drawings published in *Habitat* express this modernist vernacular sensibility, he also perceived the housing as economically and socially primitive, lacking hygienic conditions, urban infrastructure, and basic values that were necessary to bring this population out of a chronic state of underdevelopment. In his descriptions, the riverside houses appear less as a manifestation of popular knowledge than the result of the limitations imposed by precarious living conditions; therefore the urban planning of VSN could not be reduced to private commercial objectives, but had to be conceived as an instrument through which social change and progress would reach this remote region.

The architect was responsible for all phases of planning and construction: the overall calculus of the city's dimensions, infrastructural capacity, and demographic concentration; the urban layout, water, and sewage infrastructures; the architectural project of the buildings, down to the design of all the furniture. Like most architects of his generation, Bratke shared a common faith in the transformative power of architecture and urban planning, and to a great extent, VSN was as representative of this belief as the capital Brasília. Constructed concurrently, both deployed the symbolic and organizational models of modernist architecture as a total environment projecting an entire new collective subjectivity of progress and development that was central in constructing a national imaginary of modernity. Unlike the new capital, however, VSN was meant to be a closed compound, in the face of which Bratke's project for a civic utopia was at best a contradiction. Yet for the architect this was as much a potential as a limit. The company's tutelage over the life of the workers "did not incentivize the spirit for struggle," he noticed, but if the city was correctly governed, workers would get "habituated to hierarchical discipline," and providing that the presence of the company's administration was not excessively ostensive, the city could turn out to be a "school for gregarious life."[14] Bratke suggested that VSN should count on a municipal council which would slowly assume political–administrative functions while the city itself would gradually open to commercial exchange and migratory fluxes. VSN combined corporate discipline and modernist planning as pedagogical instruments through which a capitalistic ethos and frictionless social hierarchies would be produced at the urban scale. Despite its reduced dimensions, the architect argued that VSN was meant to serve as an "exemplary city": isolated by the forest, a new type of community would be progressively crafted by the harmonious designs of modernism during the forty years that the mining concession lasted.

Territorial Design

In 1971 the economist Walter H. Pawley, at that time the head of the Policy Advisory Bureau of the Food and Agriculture Organization of the United Nations (FAO), published a short article presenting the prospects for maintaining world food supply given the historically unprecedented growth of the human population. If current trends continued, "In the Year 2070"—as the text was titled—Pawley estimated that the planetary population would have increased ten-fold, totaling around 36 billion people, and generating all sorts of pressure on the earth's capacity to sustain the "human race." Pawley explained that the problem, however, was not so much the "global total," but more precisely the abnormal rates of demographic change in the "low-income/high-fertility" Third World, which by then would have ten times the population of the West. The widening gap between limited resources and ever-growing demand—a deregulation in the "land-man ratio"—was framed as a global phenomenon, yet its socio-spatial consequences would be most severely experienced in "developing countries" of Asia, Africa, and Latin America. Amid the crowds of the poor, where life was already insecure, lack of protein would lead to further poverty and ecological depletion, and ultimately to "social upheaval and political and administrative breakdown." While the inevitable "demographic problem" called for international concern, its explosive political effects were located in specific territories and populations, which in turn could become a potential threat to global stability.[15]

The economist envisaged that the dramatic rise in the demand for food could be tamed by two achievable technological breakthroughs: continuous cultivation of the humid soils in the tropics, and the desalinization of seawater for "perennial" irrigation. The latter would turn the Sahara and the Australian hinterlands into a "sea of waving green," while the former would open up vast tracts of land for technology-intensive agriculture in the equatorial forests of the Congo and the Amazon. By integrating these continental areas into agricultural production, within one hundred years the world harvest could be fifty times the global output of 1971 and all the 36 billion people would share the diet of the developed world. Despite the sci-fi tone of his speculation, the underlying message was nevertheless pragmatic: to absorb the effects of the "population bomb"—as the debate became popularized—he urged opening up new frontiers and spreading development, simultaneously enabling new resources to enter into the channels of global supply and curbing the political instability generated by mass poverty.

When Pawley suggested the agricultural colonization of the Amazon, Brazil was under the most repressive period of a twenty-year-long military rule, GDP-growth was at its peak, and the government was about to implement a full-scale project of occupation in the tropical forest that was not far from the futuristic prognostics of Pawley's proposal. Two years after the 1964 coup, the military had launched a program named "Operation Amazon," which introduced a new legislative package that converted the entire basin into a detached jurisdiction under direct control of the national government. The overlap between the natural limits of the Amazonas watershed with the artificial boundaries of a legal enclosure allowed

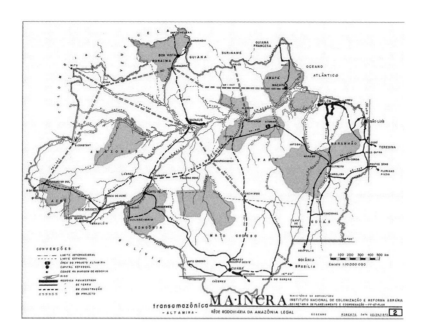

9.09 Territorial design: the "urban-matrix" proposed for the Amazon as part of the Plan of National Integration, 1971. The shaded areas show the "development poles."

the reconceptualization of "The Amazon" as a single and continuous space of intervention and projected a complete new cartography onto the area. With the financial and technical support of the United States, the military carried out unprecedented remote-sensor mapping to inventory its mineral, oil, and timber resources. The multiplicities of socio-ecological systems that inhabited the basin were reduced to a homogeneous resource-terrain for which planning was formulated and executed at territorial scale. New bureaucratic and economic channels were established in order to ease the barriers for capital investment and a dedicated administrative structure was created to oversee and promote projects of economic occupation. This legal-cartographic apparatus formed the basis for an aggressive plan of "territorial design" that would be implemented during the 1970s and the 1980s.

Departing from the northern delta of the Amazonas River, in the region where VSN is located, a road corridor was planned to run parallel with the northwestern international borders with Suriname, the Guyanas, Venezuela, Colombia, and Peru, following a semi-circular route that duplicated the limits of state sovereignty as a defensive asphalt-line imprinted on the ground. This "sanitary cordon" extended south until connecting with the arc formed by the two other major arteries planned alongside Brasília, the system altogether completing a circle around the legal/natural basin at the mouth of the Amazonas. Cutting through the middle of the forest, a major longitudinal artery—the infamous Transamazônica Highway—was planned in line with the great river, and a south–north circulation axis was designed to channel livestock and mono-crop production from the central plateaus of the country to global trade circuits via the riverine ports of Santarém and Manaus. This new transport network was regulated by an exclusive legislation that determined the expropriation of a strip of land sixty miles wide on either side of the roads wherein state-led and private programs of colonization would be implemented.

Corridors were planned to respond to multiple functions: they operated as protection barriers at the international borders, allowed heavier drainage and

BRASÍLIA
ALTAMIRA
BRASÍLIA

5 mil quilômetros
de carro pela Amazônia

circulation of material from the interior, and opened up migration routes that
would serve as structural spines for the occupation of the entire region. Areas of
potential resource exploration were translated into a series of "development
poles" arranged at strategic intersections of the infrastructural mesh. Like the
previous experiences of Fordlandia and VSN, these nodes were designed to
function as territorial islands of tax-exempted jurisdictions, legally prepared to
attract and accommodate transnational enclaves of resource extraction, extensive
cattle ranches or capital-intensive agri-business plantations. In Manaus, the major
transport hub of the upper Amazon, the government created a free-trade zone
to promote the installation of *maquiladora*-like assembly plants. Large-scale
hydroelectric dams were planned on both sides of the basin in order to meet the
energy demands of heavy mining activities and rapid urbanization. Telecom-
munication wires and antennae, electric cables, fiscal incentives and lines of credit
were to follow the roads, and further state investments in the modernization of

existing infrastructures and the construction of new ports and airports would improve logistics in the region.[16]

The rationale behind the planning scheme derived directly from the geopolitical–developmental discourse gestated in the postwar period. Elaborated and diffused by Brazil's Superior War School, a military–political think-tank created in 1949 under the influence of the US War College, the "Doctrines of National Security" enforced during the military regime merged spatial planning and infrastructural modernization with political containment and territorial control.[17] For the geopolitical strategists of Brazil, the low level of resource exploration, "demographic voids," and widespread underdevelopment made the Amazon a politically unstable and militarily fragile zone. Whereas the attempt to "complete" the consolidation of the national territory and "close the frontier" reiterated historical concerns over sovereignty of the Amazon, during the Cold War those fears were intensified by hard-line nationalism and the circulation of ideas like Pawley's, which for the most part were locally perceived as attempts to internationalize the region and subtract it from the control of the Brazilian state.[18] More importantly, these development/security strategies were directly related to fresh anxieties about "internal enemies" and concerns over a potential communist *foco* in the jungle, an environment that, from Cuba to Vietnam, had proven to be an enemy hard to defeat. The territorial plan should thus combine military strategy and economic rationales, articulating security and development via the deployment of spatial infrastructures. Echoing the imperialist genealogy of geopolitical thought, the military regime sought to transform the Amazon into a "live frontier," both productive and expansionist, in order to strengthen sovereignty over international borders and neutralize potential internal pockets of insurgency.[19]

9.11 The Amazon frontier: location of the areas and projects discussed in the text

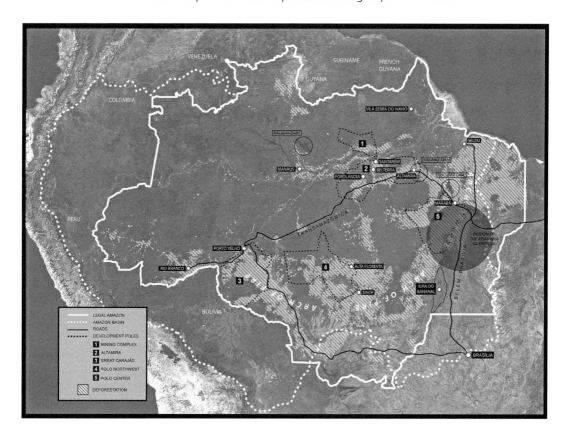

Frontier Architecture

It is impossible to understand the spatial transformations that took place in the rainforests of the Amazon without looking at the region's counter-part environment—the dry lands of the Northeast. Seasonal droughts brought by El Niño to this region had made the Amazon a historical refuge for populations escaping the harsh climate of its semi-arid zones. Waves of migrants either followed the rising prices of the forest's commodities in the global market, or were promoted by direct state intervention. In most of the situations, all these forces—environmental, economic, and political—have contributed to displacing people, as in the case of the migratory programs implemented during the Second World War, when thousands marched westward motivated by higher demand for rubber and the devastating drought of 1942. In 1970, the Northeast was hit by another severely dry period. For economists, planners, and technicians working in the government developmental agencies, the region suffered from a sort of Malthusian plague: overpopulation combined with land scarcity, water shortage, and lack of technology resulted in diminished food supply, environmental depletion, and famine. Employing the lexicon of regional planning theories and spatial-economic models, planners described the Northeast as a "depressed-periphery" in relation to the "core-regions" situated on the São Paulo–Rio de Janeiro axis. Whereas the problem of the Northeast was demographic pressure over a precarious ecology, the problem of the Amazon basin—a "non-integrated periphery"—was precisely the opposite: low demographic levels, abundance of resources, and economic underuse. The Transamazônica Highway would serve as a spatial fix for both extremes: channeling population flows toward the humid hinterlands, as a prominent geographer wrote at that time, would "alleviate the economic, social, and political tensions" in the semi-arid zones.[20] Indeed, the problem was political as much as environmental, for in the Northeast, like in the rest of the Brazilian territory, an unequal share of resources and the maintenance of archaic labor relations inherited from the colonial period had perpetuated the monopoly of a reduced class of landowners over fertile ground. Immediately before the military coup, nation-wide popular pressure for land reform was on the verge of explosion, particularly in this area, which had characteristically been a stronghold of peasant leagues and rural labor organizations. The corridors of colonization in the Amazon were designed to function as security valves to alleviate the potential of political unrest over the historically uneven distribution of land in other areas of the country.[21]

After the drought struck in 1970, the government created the National Institute of Colonization and Agrarian Reform (INCRA), which was responsible for planning the colonial settlements at the margins of the road corridors. José Geraldo Camargo, the architect who was in charge of the overall design, developed an urban mechanism based on an abstract diagram of occupation that could potentially extend throughout the road network, regardless of the specificities of each location. Three types of settlements organized according to degrees of urbanization defined a basic module of colonization. The minor cell would accommodate 100 to 300 families and elementary services (termed *agrovilas*); the medium up to 500 (*agrópolis*); the larger had a planned capacity of 20,000 and would function as a fully equipped urban center (*rurópolis*). Every major urban node served twelve or more medium-scale settlements, which served another twelve or more smaller rural communities. This module formed an urban pattern repeated at every eighty-seven miles, virtually projecting a non-stop linear city as endless as the Amazon. For each of these nuclei, the architect designed "theoretical" urban models that could be adapted according to the topographical variations of the basin. Camargo explained that the advantage of this planning scheme was that it allowed flexibility without compromising the overall "social planning" dimension of the project, since the main spatial organization was registered in the diagrams that guided the entire process. For the architect, the agrarian problem was not only the question of land distribution, but also the lack of urbanity that predominated in backward agricultural areas. Agrarian reform, therefore, should be planned as a form of urbanization, or what he called "rural-urban planning."[22] Whereas the modernist schemes of his theoretical models refer directly to the *superquadras* of Brasília, the truly modern gesture is to be found elsewhere, in the frontier character implied in the urban mechanism, which, similar to the new capital, articulates the rationality of modernist design with territorial colonization.

A pilot program was initiated at the margins of 290-mile point of the Transamazônica Highway, between the Xingu and the Tapajos Rivers, a politically sensitive area which was of central importance in the territorial planning that was being implemented in the Amazon. Named after a local urban center, Altamira-1 was strategically placed between two developmental poles: to the north a huge mining complex was under construction; to the south another was planned in the Serra dos Carajás, a mountainous region rich in iron ore around which the Karajá used to live before being confined in the Ilha do Bananal. Two hydroelectric dams were planned for that zone, which was also designated to accommodate large agricultural and livestock production farms. The enclosure mechanisms that came together with the demarcation of giant territorial jurisdictions dedicated to mining consortiums, the flooding of extensive areas,

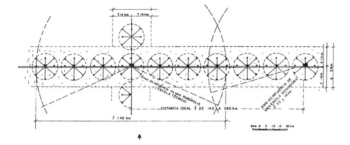

9.12 Territorial planning for colonization: the diagram that structured the occupation of the Transamazônica, with each circle representing a diameter settlement module of sixty-five miles; the project for a single module located along a hypothetical section of road

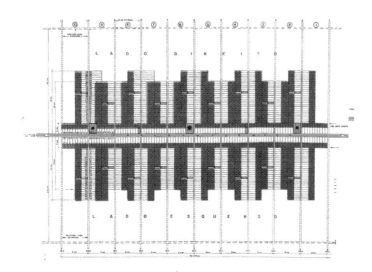

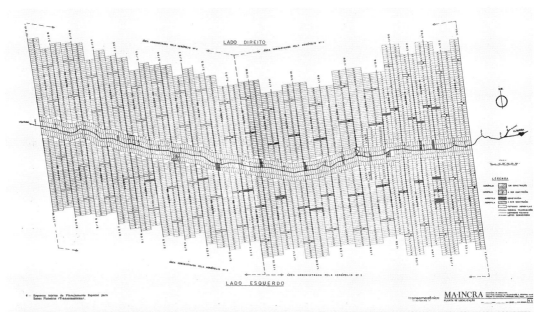

9.13 Non-stop city: despite its chaotic appearance, the "fish-spine" patterns of deforestation in the Amazon follow the original 1971 design; satellite image of the Transamazônica showing the area where Altamira-1 was implemented, 2011

and the promotion of *latifundios*—large agricultural estates—led to massive dispossession of local populations. The planned frontier triggered a chaotic process of land speculation and squatting that itself became a self-propelling engine of colonization. De facto occupation became a mechanism to claim de jure ownership to an otherwise unlegislated area, and indigenous tribes and peasant families were frequently displaced under physical coercion. Conflicts grew increasingly violent in this region from the first years of Operation Amazon, and by the early 1970s, after a guerrilla *foco* organized by the Communist Party was uncovered some 300 miles southeast of Altamira, the entire zone was put under military occupation. Very little is known about the counter-insurgency operations because the archives remain difficult to access even today, but what is certain is that the ostensive military presence was directed equally to the sixty-nine men and women who fought in the jungle and to the explosive social tensions over land distribution that pervaded the region.[23] Altamira-1 was an integral part of that overall plan of state intervention driven by the militarization of the agrarian question.[24] Following the geopolitical rationale through which the military regime framed the Amazon, the project was designed to absorb the effects of land conflicts by equating spatial planning with political containment.

As early as the mid-1970s, the guerrillas had been dismantled and the pilot program was slowly abandoned. Many minor settlements and five larger urban nodes had been constructed, but the aim of settling one hundred thousand

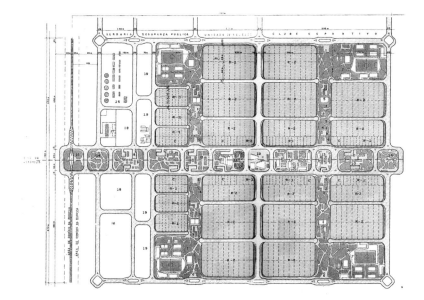

9.14 (left) Brasília's Superquadras in the Amazon: a theoretical model of a medium-sized urban settlement, 1972

9.15 (below) Altamira-1: the cover of the official report published by the National Institute of Colonization and Agrarian Reform in 1973

families was a complete failure. Apart from the fact that the soils of tropical forests are not suitable for agricultural production, speculation over land, seasonal floods and lack of technical support made life in the Transamazônica a difficult enterprise for poor peasants. Yet the basic scheme of colonization designed by José Geraldo Camargo was nonetheless successful. Over the years it has been continuously replicated throughout the road-corridors and has effectively turned into a model of occupation: although lawless and apparently chaotic, the "fish-spine" patterns of deforestation that currently shape the landscape of the Amazon precisely follow the spatial logic of the original project.

Counter-Spaces

After the abandonment of Altamira-1, state planning and financial investment were focused on further promoting large-scale capital-intensive projects, such as heavy mining activities and mechanized agriculture. Pressed by financial crisis and engulfed in foreign debt, the military regime intensified export-led production of commodities, and the role of the Amazon as the main source of primitive accumulation in the national economy increased. From the late 1970s onward, the government quickly advanced three of the developmental poles: the Great Carajás Project, an integrated mine–railroad–port system situated on the eastern side of the Amazon; Polo-Center, located in the south-center area of the basin, where the government opened up vast tracts of land for private colonization and stimulated the creation of the core-region of Brazilian soybean production; and Polo-Northwest, to the western side of the basin, wherein other rural settlements were implemented, and which had a similar fate to the chaotic occupation experienced in the region of Altamira-1. Observed together, these three poles stand at the origin of what today is known as "the arch of fire," a lawless slash-and-burn frontier line that increasingly threatens the remaining forest of the upper basin.

9.16 Satellite image showing the changes in land-use of the Polo-Northwest, 1975 and 2001: the fish-spine pattern of deforestation was converted into a "vernacular" mode of occupation in the Amazon

Contrary to usual reasoning, the patterns of destruction we now see in the Amazon are not the outcome of lack of governmental control over the territory, but the direct result of a well-orchestrated planning scheme. Its basic blueprint was aimed at transforming spaces that were practically outside the circuits of global

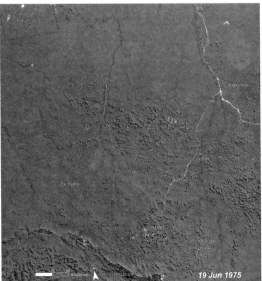

19 Jun 1975   19 Sep 2001

capitalism into a resource-frontier designed to operate at the scale of the world-market. Similar to other frontier-zones of the Third World, the supply-channels carved out of the Amazon were shaped by the ideological polarization between capitalism and communism that divided the political spectrum of the time, their modern designs were formulated and legitimized by the discourse of development that dominated the language of planners and politicians, and they only became possible through the articulation between the financial support of international banks, the expansive power of transnational corporations, and the centralized planning apparatus of a militarized state. While Brazil experienced unprecedented economic growth during that period, poverty in fact increased, following a wider gap in wealth distribution and an impressive movement toward land enclosure and deforestation. Despite its destructive effects on the ground, however, the project of territorial design put forward in the Amazon was a huge success for state propaganda. Images of new roads and cities being built from scratch were widely diffused throughout the media and "the conquest of the Amazon" came to symbolize the march of a modern nation toward a developed future, serving as an ideological space through which the military regime built its political legitimacy, particularly in the early 1970s, in the period of harshest repression.

The same ideological underpinnings continued under the "New Republic" after the process of re-democratization in 1984. Five years later, when Brazil was about to hold democratic elections after two decades of dictatorial rule, the frontier was completely out of control and the destruction of the Amazon became a matter of international concern. Images of the forest being torn down, which used to appear as signs of modernization, increasingly became perceived as one of the most expressive examples of a process of environmental depletion that by that time had already reached planetary proportions. The year 1989 was peculiar in our history not only because we witnessed the collapse of the bipolar world order of the Cold War, but also because the collapse of our ecosystem became an unavoidable political and ethical issue at the global scale. While environmental discourse began to infiltrate the sensibility of the general public, the term "sustainability" started to appear as a qualifier next to "development" in the official grammar of governmental bureaucracies.[25] But most relevant were the political effects that the stream of ecological-dissent had on the ground, manifested by what historians have called the "environmentalism of the poor,"[26] a wave of localized social movements around the globe which, to a large extent, formed a truly new internationalist political alliance, for although spatially dispersed, they shared a common space of struggle—the environment. At the frontier-spaces of the Third World, ecological forms of livelihood which had been excluded as backward, pre-modern, or underdeveloped entered into the political arena to contest the epistemological apparatus that grounded the cultural, scientific, and technological pillars of a homogenized world on the path of "complete modernity"—and ultimately, potential self-destruction. By that time the Amazon basin was literally converted into a battleground: in Polo-Northwest, the rubber-tappers movement led by the political activist Chico Mendes resisted land-grabbers and forced the World Bank to withdraw financial support for the program of agricultural colonization; up north, near the border with Venezuela, ethnographers, activists, and indigenous leaders gathered to stop the construction of a "sanitary cordon" through the territory of the Yanomami; and around Altamira-1, indigenous peoples and peasant families inhabiting the margins of the Xingu River also managed to halt the construction of a large hydroelectric dam.[27]

Popular claims for a mode of relation with the environment which was not based on top-down economic rationalities and development planning schemes resisted the symbolic and territorial enclosures of the national-development project,

9.17 Altamira Meeting, 1989: as Brazil was about to hold its first democratic elections in twenty-five years, local indigenous and peasant communities called on government representatives for a meeting in Altamira to contest the construction of the first of five large hydroelectric dams projected for the Xingu River. They successfully managed to halt this project, but controversy over the construction of the renamed Belo Monte Dam continues up to the present

reclaiming the construction of common spaces whose use and management should be as plural as the diverse patterns of inhabitation that populate the forest. The modern frontiers in the Amazon constituted above all lines of conflict, but they were also channels of intense cultural exchange and political creativity. Frontiers are territories wherein a multiplicity of times and spaces come forcefully into friction, and although for the most part this encounter assumes violent forms, it also holds the potential of opening up an arena through which new political horizons are made possible. For politics, as philosopher Jacques Rancière beautifully wrote, "is not made up of power relationships; it is made up of relationships between worlds,"[28] as it was in the Amazon, when dissident ecologies came into conflict and triggered an imaginative field that pointed toward the decolonization of the modern state apparatus and its spatial articulations, generating an entire new political terrain that contested the grounds upon which modernity was built. Indeed, that is precisely the space in which we now stand.

Notes

1   This chapter is part of a larger research project that has benefited from the support of the Ministry of Education of Brasil (CAPES). I would like to thank architects Marina Rago and Luisa Amoroso for their help with the archival research.
2   In his recollections about the construction of Brasília, President Juscelino Kubitschek wrote: "I have seen Brazil from above inside an aircraft and could feel the problem and all its complex implications. Two thirds of the national territory was virgin of human presence. They were the 'demographic voids' that the sociologists talked about. The great challenge of our History was there: to dislocate the axis of national development." Juscelino Kubitschek, *Porque construí Brasília* (Rio de Janeiro: Bloch Editores, 1975), 7.
3   Lucio Costa, *Memorial do Plano Piloto de Brasília*, 1957. The idea of creating a capital city in the middle of the Brazilian territory emerged during the colonial rule of Marques de Pombal in the mid-eighteenth century. The first republican constitution of 1891 already prescribed the construction of the new capital in the central plateaus of the country.

4   The articulation between modernist aesthetics and nationalism was a central feature
    of hard-line nationalist movements of fascistic inclination as well as of leftwing-oriented
    political experiments, most notably Oswald de Andrade's "Anthropophagic Manifesto"
    (1928).

5   Marcos Felipe de Brum Lopes, *Mario Baldi: experiências fotográficas e a trajetória do
    "reporter perfeito"* (Master's thesis, Universidade Federal Fluminense, Rio de Janeiro,
    2010).

6   Mario Baldi, "A cerâmica dos índios Carajás," *Habitat* 34 (September 1956). The other
    two essays were "Dança dos Índios, no Mato Grosso" and "Costume dos Índios,"
    published respectively in issues 33 and 37, both from 1956.

7   To a large extent the political–spatial model promoted by the SPI inherited forms of
    territorial administration and political tutelage from the missionary settlements of the
    colonial period. See Antonio Carlos de Souza Lima, *Sobre indigenismo, autoritarismo, e
    nacionalidade: considerações sobre a constituição do discurso e da prática da proteção
    fraternal no Brasil* (PhD diss., Museu Nacional, Rio de Janeiro, 1987).

8   Starting with Vargas's government, expeditions to occupy the interior of the country
    grew in size, extension, and publicity. In 1943, he created a dedicated agency named
    Fundação Brasil Central (FBC, or Central Brazil Foundation) to take charge of incursions
    into the central plateaus. The most famous was the Roncador–Xingu expedition that
    led to the creation of the Indigenous Park of Xingu in 1961, the first major indigenous
    territory legally established in Brazil.

9   Greg Gradin, *Fordlandia: The Rise and Fall of Henry Ford's Forgotten Jungle City* (New
    York: Metropolitan Books, 2009), and also Elizabeth Esch, "Whitened and Enlightened:
    The Ford Motor Company and Racial Engineering in the Brazilian Amazon," in *Company
    Towns in the Americas: Landscape, Power, and Working-Class Communities*, ed. Oliver J.
    Dinius and Angela Vergara (Athens, GA: University of Georgia Press, 2011).

10 Oswaldo Bratke, as quoted in Benjamin Ribeiro, *A Vila Serra do Navio: Comunidade
    urbana na selva amazônica* (São Paulo: Pini, 1982), 10. This book reproduces long
    passages of the architect's report to the Bethlehem/ICOMI consortium. It also
    reproduces quotes from a report produced in the 1980s on living conditions in the city,
    wherein it is acknowledged that "the counterpart of high levels of the infrastructural
    development" of VSN "is the ostensive presence of the company in the life of the
    workers." Ribeiro, *A Vila Serra do Navio*, 100.

11 Judith Stein, *Running Steel, Running America: Race, Economic Policy, and the Decline of
    Liberalism* (Chapel Hill: University of North Carolina Press, 1998).

12 See Hugo Segawa, "Oswaldo Bratke: Vila Serra do Navio e Vila Amazonas," in *Textos
    fundamentais sobre história da arquitetura moderna brasileira*, vol. 2, ed. Abilio Guerra
    (São Paulo: Romano Guerra, 2008).

13 Bernard Rudofsky, *Architecture without Architects* (New York: Museum of Modern Art,
    1964), 110.

14 Oswaldo Bratke, as quoted in Ribeiro, *A Vila Serra do Navio*, 36.

15 Walter H. Pawley, "In the Year 2070: Thinking Now about the Next Century Has Become
    Imperative," in *Ceres* (FAO Review) 4, no. 4 (July–August 1971): 22–27.

16 The description of "Operation Amazon" and the plans that followed appear in many
    reports published by SUDAM, the developmental agency in charge of the overall
    planning, such as: *A Amazônia* (1973); *Amazônia: novo universo* (1975); *II Plano de
    Desenvolvimento da Amazônia* (Brasília: Minter/Sudam, 1975).

17 The most important manifesto informing the colonization of the Amazon during the
    military regime was written by General Golbery do Couto e Silva, a leading intellectual
    of the Superior War School and one of the central ideologies of the coup. Golbery do
    Couto e Silva, *Geopolítica do Brasil* (Rio de Janeiro: José Olimpio, 1967).

18 Projects like the five "Great Lakes" proposed by US strategist Herman Khan for the
    middle of the Amazon basin, or the creation of a UNESCO-led research institution for
    the Amazon in the 1940s, were often criticized by sectors of the intellectual and political
    elite.

19 One of the most lucid analyses of the geopolitical discourse/practice of the military
    was written by Francisco de Oliveira, "A Reconquista da Amazônia," *Novos Estudos* 38
    (March 1994).

20   Bertha Becker, "Crescimento Econômico e Estrutura Espacial do Brasil" (1972), in *Geopolitica da Amazonia: a nova fronteira de recursos* (Rio de Janeiro: Zahar Editores, 1982). Much of this discourse was informed by regional planning schemes and "center–periphery" spatial models derived from economic theories of development/ modernization. More generally they echoed the discourse of the regime, notably summarized by President General Emílio Garrastazu Médici who, after visiting the drought zones in the 1970s, described the Amazon with the Zionist slogan "a land without a people for a people without a land."

21   Sociologist Otavio Ianni called the program of land distribution implemented by the military a "counter agrarian reform," for its true aim was to maintain the concentrated ownership of land—the *latifundio*—rather than restructuring the system altogether. Otavio Ianni, *Ditadura e Agricultura: o desenvolvimento do capitalismo na Amazônia* (Rio de Janeiro: Civilização Brasiliera, 1979).

22   Jose Geraldo da Cunha Camargo, *Urbanismo Rural* (Brasília: INCRA, 1973).

23   Maria Helena Moreira Alves, *Estado e Oposição no Brasil (1964–1984)* (São Paulo: Vozes, 1984).

24   On the "militarization" of the land reform efforts see José de Souza Martin, *A militarização da questão agrária* (São Paulo: Vozes, 1984).

25   Most notably in the United Nations' report on the environment "Our Common Future" published in 1987, wherein for the first time "sustainability" appeared alongside the overused term "development."

26   See Rachamandra Guha, *Environmentalism: A Global History* (London: Longman, 2000).

27   Those conflicts were summarized by one of the main protagonists, the anthropologist Bruce Albert, "Indian Lands, Environmental Policy, and Military Geopolitics in the Development of the Brazilian Amazon: The Case of the Yanomami," *Development and Change* 23, no. 1 (January 1992): 35–70.

28   Jacques Rancière, *Disagreement* (Minneapolis: University of Minnesota Press, 1999), 42.

Part III

MEDIATED TERRITORIES

10

REFLECTIONS OF THE "COLONIAL"

Between Mexico and *Californiano*

Cristina López Uribe

When reading histories of early Mexican modernist architecture, one constantly finds negative references to neocolonial architecture, a style that borrowed forms in an attempt to evoke images of the Spanish colonial past. However, one cannot help wondering whether the relationship between the two has been over-simplified: in the construction of Latin American identities, modes of expression that reinforced nationalism through familiar images (such as the neocolonial) were as modern as those more in tune with the technological impulse (such as modernism) that championed an internationalist abstraction.

By the 1940s, the urban landscape of Mexico City was largely defined by neocolonial buildings, as were many other Latin American cities. Neocolonial architecture used forms associated with the Spanish colonial period not merely in a nostalgic revivalist spirit, but also sometimes reframed them within a modern conception; for example, adopting new patterns of internal distribution, and using industrially produced materials. The neocolonial had several variants, some quite accurate in reproducing the colonial era, while others only employed its ornament to ease the harsh look of utilitarian buildings. It is not the intention of this chapter to identify or classify all these variants, which would certainly be a difficult task. Instead I will focus on the most popular, known as *Colonial Californiano*, which was used widely in the 1930s and 1940s for new residential developments in Mexican cities.

Despite the number of buildings made in this fashion, few studies have been done of them. Criticism and historiography has largely considered them from the point of view of the Modern Movement, which judged the development of this architecture as a misstep. The two were framed as exact opposites: the neocolonial's superfluous ornamentation and expressiveness versus modernism's low cost and engagement with industry. Rhetorically, modernism based its legitimacy on its struggle against neocolonial architecture. Sometimes the defenders of modernism had solid grounds to believe it was under siege, as in 1936 when Mexico City's local authorities tried to favor neocolonial over modernist construction, but this was an isolated incident.[1] I would argue that the main reason for modernist animosity was that its aesthetics did not achieve popular acceptance as early as the neocolonial did.

It is important to consider that nationalism undergirded neocolonial architecture. Around 1920, at the end of a decade of violence following the Mexican Revolution, intellectuals gathered at El Ateneo de la Juventud argued that the nation needed to reconstruct its identity, with a "New Mexican architecture" that should be related to that of the colonial period.[2] For them, the seventeenth and the eighteenth centuries represented a mythical moment of racial and cultural mixture

produced by blending indigenous and European elements; colonial architecture of this period was claimed as the evidence of this new cultural mixture, a truly Mexican product. As such it was considered to be the best source for the creation of a new national style suitable for state-promoted architecture. Architects graduating in the 1920s from Mexico's only architecture school, the Academia de San Carlos, used this "national style" on many of their projects in an attempt to valorize the Mexican over the prevailing French academic taste. In this context, neocolonial architecture was used to represent the country in international events, such as the Exposição do Centenario do Brasil in Rio de Janeiro in 1922, where the Mexican pavilion was designed by Carlos Obregón Santacilia. As the neocolonial was similarly developed in other Latin American countries, it was also used to represent the Latin American region, and even, in some cases, the whole American continent, as in the Pan American Exposition held in 1901 in Buffalo, and the Panama-Pacific International Exhibition held from 1914 to 1917 in San Diego. This variant of neocolonial architecture claimed to be historically correct in its exterior appearance—it looked very similar to colonial-era buildings—but in plan and section it was a recognizable product of the École des Beaux-Arts.

However, the real threat to modernist aesthetics was the mass-appeal of the neocolonial, as evidenced in the residential architecture that appeared from the 1930s to the 1950s—work produced not only by architects but also by real-estate developers, builders, and engineers. Here the influences were different, primarily mass-media images, created not in Mexico, but the United States. Although it also made use of the colonial aesthetic, this architecture did not have the same solid ideological support, and the guild of architects considered it to be shallow and commercial. Nevertheless, it enjoyed a high degree of popular acceptance due in some part to nationalist sentiments, and also because it expressed the desire of the emerging middle classes to be in tune with the modern life projected in movies and other media from the United States.

Thus we are facing two different architectures. On the one hand, a few examples promoted by the academy, supported by well-consolidated theoretical precepts, justified by nationalistic ideology and the state, and used primarily in public buildings. On the other hand, the most common residential examples, hundreds of buildings inspired by mass-media images from magazines and movies, and strongly despised—at least in their discourse—by modernist architects and modernist historiography. I have selected this domestic architecture as the focus of this chapter because it seems that thinking about the two variants as if they were the same has prevented an understanding of the latter. I will try to trace how the images of the colonial that influenced this architecture were created— first of all in the minds of Southern California architects; how it was represented in the mass media; and how Mexican architects adopted and adapted these images to their own locations in Mexico. In this sense, I aim to explore what has been described as a "strange play of mirrors, where the United States was virtually copying Mexico, which in turn copied from the United States what had been copied from it."[3]

Californian Views

In 1884, Helen Hunt Jackson's romantic novel *Ramona*, set in Southern California, had an immense commercial success. The reasons for this were several. Until 1848, what is today California was part of Mexico. With its newly gained territories, the United States naturally sought to expand its economic growth through railroad construction to reach the new Western states. This expansion also triggered tourism, as ordinary people felt curious about the new US territories. The appeal of *Ramona*, and later of the first Hollywood movies, was due to their representations of this unknown world: California provided US audiences with scenery that had the double advantage of being both exotic and its own.

The trail that I am looking for begins here with architectural styles developed in Southern California: Spanish Colonial Revival and Mission style. Well-equipped modern houses built in Colonial Revival styles that celebrated the era of New England settlement were popular all over the United States, but on the nation's new frontiers where the heritage was not English but Spanish, the term "colonial" took on a different meaning and an entirely different look. These Spanish-influenced buildings had a special charm, for they not only had roots in history, but also conjured up romantic images of a far away, long-ago, and idealized Mediterranean countryside, a desirable destination no longer available with the arrival of the First World War.

10.02 Ramona Candy Company bon-bon wrapper incorporating the character of Ramona and images of colonial architecture

However, the image of Mexico itself also played an important role. Even in 1910, California was still perceived as a land of Mexicans, and of the frontier; in many ways, the border between the US Southwest and Mexico has remained imprecise. The colonial architecture of this region was that of the evangelizing missions and some *ranchos* or *haciendas*, within the desert landscape where the early Hollywood movies were filmed. The new Californian identity that was created was "exotic" and "Spanish"—but it could not be seen as Mexican. This territory had been won from Mexico in a war, and for this reason had to have a character of its own. Furthermore, images of the Mexican Revolution circulated widely in the United States during the 1910s—no war had ever been filmed and photographed as extensively—and had created both an eager following of the events of the war and a negative image of the country.[4] Covered on a daily basis, hours and hours of footage (some of it simulated) was taken for the newsreels and for the first feature films, which US audiences understood as "Westerns." As Margarita de Orellana has observed:

> From the outset of the Mexican Revolution, [US] cinema producers sought to capitalize on the great public interest shown in its progress. While the distinctive vegetation, the architecture, and the local customs were already box-office attractions, the interest grew now that there were battles between rudimentary armies, led by picturesque and very dramatic leaders.[5]

This increased the US public's fascination with the perceived exoticism of Mexico, but this kind of coverage also ensured that up until the 1930s mass audiences in the United States saw Mexico as a violent land of bandits, out-laws, uneducated, uncivilized, and strange people.[6] Because of this prevailing negative image of all things Mexican, the architectural styles that we are referring to were given the name "Spanish" in order to erase the term "Mexican."

The new Californian architectural styles recovered Spanish colonial forms, but used a modern rational plan influenced by the Arts and Crafts movement, then an important component of home furnishings and interiors throughout the United States, seen in Californian architecture with the work of John Galen Howard, Greene and Greene, and Bernard Maybeck.[7] The houses "looked" Spanish or Italian, but they also unwillingly looked Mexican, an association clearly not openly acknowledged. Some of the best examples are George Washington Smith's Dracaena House (1920), which clearly recalls a traditional Mexican small-town house.

After the violent years of the revolution, the image of Mexico in the United States would start to change. The country came to be seen as an exotic, idyllic, and timeless rural locale, apart from the modern, industry-dominated world, where essentially "American" values—different from the cultural dominance of Europe—could be found.[8] Examples from popular and commercial culture reveal in a less self-conscious manner the prejudices and expectations of mass audiences. The exotic and sensual travel posters and postcards, and also the brochures produced by railroad companies—images designed to lure tourists south—contributed to the success of the "Spanish" styles. The first tourist accommodations south of the border, as well as train stations, casinos, and beach hotels along the railroads, were designed in these new, fashionable styles.

However, the wide acceptance of this architecture in California was not enough to complete the phenomenon I am tracing. The development of the movie industry and its "star" system, taking place precisely there, in Hollywood, helped these images to cross the border. Some of the first Hollywood features were Westerns set in California or Mexico, and some of the early movie stars played

10.03 (above) George
Washington Smith,
Dracaena House, Santa
Barbara, California, 1920
(Reprinted with permission
from Architectural Record
© 1920, The McGraw-Hill
Companies, www.
architecturalrecord.com)

10.04 (left) Douglas
Fairbanks in *The Mark
of Zorro*, 1920

Mexican—or almost Mexican—characters, the most famous being Douglas
Fairbanks, who played Zorro in *The Mark of Zorro* (1920), disseminating a particular
image of Mexico around the world. This movie was extremely successful in Mexico,
where people are particularly fond of masked heroes. In October 1934, Fairbanks
was the guest of honor of Mexican president Abelardo Rodríguez (1932–1934) at
the inauguration of the Palacio de Bellas Artes (National Theater). In 1932, the
president had commissioned Rodolfo M. Fernández, who had studied architecture

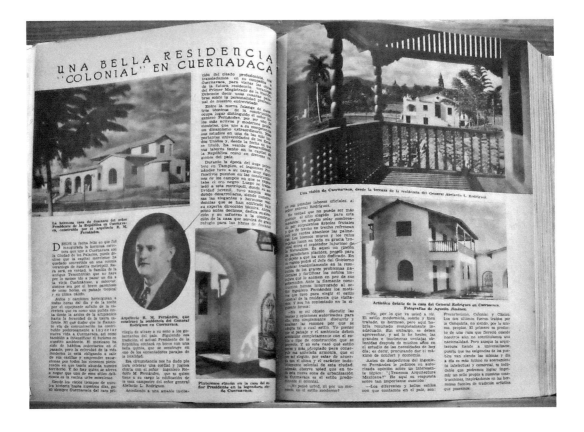

10.05 "A beautiful 'colonial' residence in Cuernavaca"— according to the magazine *Revista de revistas*. R. M. Fernández, General Abelardo L. Rodríguez Residence, Cuernavaca, Morelos

in the United States, to build him a weekend home in the city of Cuernavaca using the Spanish Colonial Revival style, just like the houses of Charlie Chaplin and Douglas Fairbanks on Hollywood's Bonvue Avenue. Being associated with the cinema—the latest symbol of modernity—gave the neocolonial an irresistible charm for the new and growing Mexican middle class that was anxious to keep up to date with world trends.

Architects such as Fernández who returned to Mexico with a taste for the neocolonial developed in the United States, built houses in all the most important Mexican cities; in addition, US architectural magazines such as *The Architectural Record*, *Architectural Forum*, *Architectural Digest*, *Architecture*, *Pencil Points*, and *The American Architect*, were widely read all over Mexico. Unlike the magazines promoting the European avant-gardes, these publications were within easy reach for many Mexicans. Images of the neocolonial—in features on projects by architects of the Southwest and in advertisements for construction materials— were widely published in the magazines, and somewhat ironically proved to be useful sources for Mexican architects and builders searching for "local expression" via these imported images.

However, while Mexican colonial houses of the seventeenth and eighteenth centuries fronted the street and were organized around a central courtyard, the new houses of the *Colonial Californiano* were organized around a central hall and set in the middle of the plot, liberating the garden around it, in tune with the ideas of the garden city. Thus Mexican neocolonial houses of the 1930s and 1940s, whether of *Colonial Californiano* or almost any of its variants, had much more in common with models from the United States than with the Mexican colonial architecture just around the corner; the enthusiasm for these models was so great that original colonial houses were even demolished to make way for new neocolonial ones. The US influence on the Mexican neocolonial was also evident in the naming of the elements of the domestic program, many of which were innovations in the Mexican context: the hall, the porch, the closet (in English in the original plans), as well as the pantry (*despensa* or *alacena*), the sewing-room (*costurero*), and the breakfast room (*desayunador*).[9] None of these spaces had previously existed in Mexican residential layouts. In this way, the modernization of the internal distribution of the home appeared first in neocolonial residences, not Functionalist ones.[10]

But the Mexican examples were not merely copies, and some new variants soon appeared as the neocolonial became acclimatized to its new location. The best example is the so-called neo-baroque, which was much in demand by the well-off classes, celebrities, politicians, and local movies stars of the late 1930s and

A COUNTRY HOUSE IN CALIFORNIA

Photo. W. M. Clarke

HOUSE FOR CYRUS PIERCE, LA QUINTA, CALIFORNIA

GORDON B. KAUFMANN
ARCHITECT

10.06 "A country house in California." Gordon B. Kaufmann, Cyrus Pierce House, La Quinta, California, 1930 (Reprinted with permission from Architectural Record © 1930, The McGraw-Hill Companies. www.architecturalrecord.com)

10.07 "Construcciones
Modernas" building
company, example of a
Spanish-style residence,
Cemento, 1929

PRIMOROSO MODELO DE RESIDENCIA ESTILO ESPAÑOL

1940s. These buildings used references to baroque ecclesiastical architecture and medieval European castles on their exteriors, but featured modern, well-equipped interiors organized around the central hall. It is true that these architectural styles were the aesthetic choices of a conservative social class, which was not ready to embrace modernist aesthetics and instead flattered itself with the aristocratic allure of this architectural ornamentation. But it is important to state that this architecture represented for them the glamour of the modern world and the "American" way of life, with the added benefit that it was an image of modernity (adopted by the latest movie stars) that was somehow originally inspired by the Mexican past. Paradoxically, it nourished their nationalism. Both of the Californian creations—the movies and the "Spanish" neocolonial architecture—proved to be sources for the invention of a modern Mexican identity.

In 1927, the Mexican art magazine *Forma* published an article discussing Antonio Ruiz's drawings of colonial interiors, which were to serve as designs for Hollywood sets.[11] Having studied architecture for many years, in 1925 Ruiz had traveled to

Núm. 26
Noviembre de 1928

CEMENTO

10.08 A colonial patio used to promote concrete, drawing by Jorge González Camarena, *Cemento*, 1928

10.09 Antonio Ruiz, sketch for a film set, *Forma*, 1927

Los Angeles to work as a set designer at Universal Studios. The article explained that some Hollywood art directors showed an interest in his drawings because "there was truth and beauty in them," but they felt that to ensure commercial success the projects needed to be enlivened by "what the US public knows as the 'Spanish of Mexico' [in English in original], that is to say, a falsification of traditional representations which emphasizes praise for the exotic exterior."[12]

Looking at the drawings, one can see that the architecture that Ruiz represented was the one he was familiar with from Mexico: urban colonial baroque architecture of the seventeenth and eighteenth centuries. The article continued, almost congratulating Ruiz for not being a greater success in the movie industry: "Only a strong calling as an architect, combined with a painter's vision, could successfully handle the responsibilities of 'colonialism' in these times." Around this time, architects were debating the "correct" usage of the colonial—a topic that was on the agenda at the Third Pan-American Congress of Architects in 1927: "The Colonial Style, Origin, Characteristics, and Modern Application."[13] Nonetheless it was obvious that Hollywood movie sets would not have constituted an approved usage according to academic architects.

In Ruiz's drawings we see no desert landscapes, no cactus or *maguey*, no peasants, and no Mexican *señoritas*. Maybe that's what art directors meant by the "exotic exterior" needed to enliven his scenes. In fact, the only exteriors Ruiz shows are the traditional patios common in this type of building. These examples reveal the different images of the colonial envisaged by a cosmopolitan Mexican architect on one side, and a Hollywood art director on the other. The anonymous author of this article laments the set-influenced architecture of Los Angeles, but even more the fact that Mexican architects were trying to emulate it—and this was only 1927, when the influence was not yet at its height. In the end, Antonio Ruiz devoted himself to painting, and became one of the most important chroniclers of Mexico City's urban folklore. Perhaps ironically, he was also very influential on the Functionalist architect Juan O'Gorman, whom he met in 1920: according to O'Gorman, Ruiz taught him the painter's trade.[14]

10.10 Antonio Ruiz, sketch for a film set, *Forma*, 1927

Mexican Reflections

A brief historiographical analysis demonstrates that histories of Mexican architecture have found it difficult to place the neocolonial within a linear trajectory that assumes a teleological evolution toward a correct modernist architecture. Among the descriptions used to explain this phenomenon are: "architectural absurdities," a style of "good intentions but lean results," a "false" style, a "survival that forms part of Eclecticism," and so on. All these descriptions depart from two premises: first, that copying past styles is not valid; and second, that the search for a national architecture is relevant, but not through this route.

As I mentioned earlier, modernist architecture based its legitimacy on attacking the neocolonial, since it was much closer to popular taste. The strongest argument in favor of modernism was its economy, and therefore its suitability in fulfilling urgent social needs. However, even the first buildings produced by the post-revolutionary state were in a neocolonial idiom, which was less alienating for people than the sobriety of modernist architecture. Other arguments, concerning modernism's employment of new materials and construction technologies, proved to be less powerful because neocolonial architecture had made use of the same innovations: advertisements for cement initially used images of neocolonial houses; later, the neocolonial would appear simultaneously with modernist architecture.

According to Federico Sánchez Fogarty, publicist of the Tolteca Cement Company —one of the strongest promoters of European-influenced Functionalist architecture in Mexico—by the beginning of the 1930s neocolonial and Art Deco architecture no longer served to represent modernity. Instead, he promoted—to the benefit of his company—what he called "the aesthetics of concrete": Functionalism. By the 1930s, many influential intellectuals and artists felt more in tune with Functionalist architecture, as did the government of Mexico, which strongly supported it for state-promoted projects instead of the expensive neocolonial style of the 1920s. Functionalism was also presented as a more practical way to fulfill the social needs and the modernization of the country, and its appearance of austerity and truth to materials appealed to the political ideals of those years. Almost all of the articles Sánchez Fogarty wrote for his magazine *Tolteca* (1928–1932) promoting Functionalist and modernist architecture also included a critique of the neocolonial. Within the government, the battle was already won, but there was still a long way to go with the general public. His comments tended to be exaggerated, such as the 1931 article "The Farce of Los Angeles" written after a trip he had made precisely to get to know what he called "that false *Hollywoodesque* style," expressing the hope that *Tolteca* could help to "slow the advance of the false Angelino style."[15] In a subsequent article, he explained:

> My increasing aversions ended up being focused on the so-called "Colonial Spanish" architecture . . . it was in . . . Hollywood, where that architecture of little white lies incubated, and later was transplanted—it seems incredible!— to an entire country like Mexico, even though it is in Mexico where we should detest it the most, as we detest the *jotas* danced by *charros* [tawdry horsemen], to the beat of a poorly played tango.

> The only thing I regret is . . . that I have never been capable of writing in favor of European Functionalism with the same eloquence that I used in opposition to the touristic and merely theatrical aesthetics of Southern California.[16]

Ironically, it is thanks to the rage of the defenders of modernist architecture such as Sánchez Fogarty that we have some contemporary writings dealing with neocolonial architecture.

The first book on modern Mexican architecture, compiled by the US photographer Esther Born, included a couple of neocolonial examples among the modernist majority, but this was made from an outsider's point of view.[17] In his important historical outline published in 1950,[18] José Villagrán García clearly took advantage of the opportunity to attack "anachronistic" styles, as he called them ("anachronistic-exotic" in the case of Eclecticism, and "anachronistic-national" in the case of neocolonial), in order to substantiate the argument for modernist architecture. This is also the case of Carlos Obregón Santacilia, one of the chief proponents of the neocolonial architecture of the 1920s, who, in his 1952 book, takes good care to distinguish his own work from other more vulgar variants.[19] He presents his early buildings as a necessary mistake that illuminates the correct path to a national architectural expression. At the same time, however, Obregón Santacilia cannot refrain from spending a little too much time describing the other "incorrect" neocolonial tendency:

> The Californian style that, at one time, was making the newly rich swoon, was imposed by some real estate developers. . . . It is the most hybrid and shameful style that has been used in Mexico; it comes from the Mexican architecture of the *haciendas* and from the villages, it went to the south[western] United States, became fashionable in California, and was mixed in with what they used there of Italian and Spanish styles . . . from there it returned to Mexico and the *pochos* ["Americanized" Mexicans] have devoted themselves to copying it; only it is worse done than there, and they were copying—fortunately I believe that this has stopped—from magazines instead of making a trip to the villages of Mexico and coming face-to-face with the real thing.

He even goes so far as to classify two variants of this architecture that he despises so much:

> the Polanco or Intestinal or Syrian-Vaseline, is a Californian made with a lot of money and more imagination, pink stone, escutcheons, and inscriptions, which try to impress the friends of those who inhabit the houses of this style, up to convincing them of their fabulous or fair economic situation.

> The Narvarte is the trashy Polanco, done with little money and cheap materials that do not last, houses done in series . . . in them bloom in caricature the niche, the lantern, the tower, the dome, the grille-work, the gargoyle; they are very cheap, and very tiny with a garage in which the car does not fit . . . but with payment on credit with easy installments.[20]

The problem continued a decade later when Israel Katzman—perhaps the most important historian of Mexican modern architecture—tried to banish all those architectural expressions that obstructed the perfect path toward modernism by labeling them under a homogenizing category: "Nationalism."[21] Accordingly—and regardless of the dates of the selected examples—he managed to mix neocolonial examples with the modernist "Mexicanizing" architecture of the 1950s, and even Neo-Aztec examples contemporaneous to modernist architecture.

In historical narratives of the 1970s and 1980s, the prevailing tendency was to understand the stages of architectural history as almost mechanically generated by political and economic events. Neocolonial architecture of the 1920s was fitted into this approach by framing it as an example of the nationalistic ideas of the Mexican ruling classes in the post-revolutionary state. But this led to omitting variants that did not fit this explanation, or when mentioning them at all, characterizing them as a despicable expression of conservative, "Americanized," middle-class Mexicans, as a symbol of Mexico's neocolonial economic dependence on the United States.

In every case, the explanation of why many of the most famous modernist architects carried out explorations in the neocolonial language has been left aside. The clearest example is Luis Barragán, who was first noticed in the United States for his neocolonial architecture, clearly influenced by the Californian styles, before being consecrated as a modernist.[22] In the 1980s, this early stage of his career would be characterized as an exploration of Mediterranean architecture—a less shameful influence—on his path toward a modernist, truly Mexican architecture. Other examples have not been studied at all, such as the radical Functionalist Álvaro Aburto, who after his brave defense of Functional architecture in 1933 ended up designing rural houses that were far from the image of modernity.[23] On the other hand, the neocolonial works of Enrique del Moral or José Villagrán García—architects of the purest Mexican modernism—have often been concealed by critics and researchers in order to present them as irreproachable modernists. There is also the case of Francisco Serrano, who made simultaneous explorations in several languages: modernist, neocolonial, *Colonial Californiano*, and Art Deco. In all his works, he proved himself a talented architect, but only recently has his engagement with the "commercial" and the "superfluous" stopped being questioned. Finally, the Martínez Negrete brothers and Ignacio Díaz Morales, prominent figures included in Esther Born's 1937 book, were subsequently marginalized in historical accounts, probably due to their neocolonial work.[24] In fact, many renowned modernist Mexican architects designed neocolonial buildings, but none of them could say so out loud.

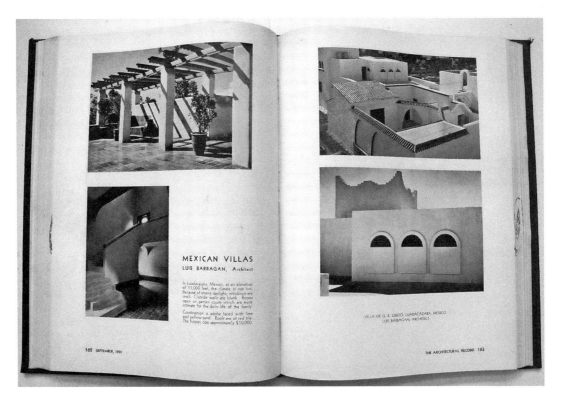

10.11 "Mexican Villas" published in *Architectural Record*, Luis Barragán, Villa of G.R. Cristo, Luadacazara, Guadalajara, 1929 (Reprinted with permission from Architectural Record © 1931, The McGraw-Hill Companies. www.architecturalrecord.com)

Conclusion

In 1901, Sylvester Baxter's lavishly illustrated *Spanish-Colonial Architecture in Mexico* introduced readers in the United States to the rich heritage of baroque buildings in Mexico, being the first and most important academic study on this kind of architecture for many years.[25] In 1930, Richard Garrison's *Mexican Houses* disseminated quite a different colonial Mexican architecture in the United States: not grand monuments—as in Baxter's book—but ordinary, anonymous, rural buildings.[26] The colonial architecture of small towns was the focus instead of the most prominent urban buildings emulated by the officially—and professionally— sanctioned Mexican neocolonial architecture of the 1920s. Ironically, the descriptions of the finishes and colors of the houses resemble those of Mexican Functionalist residences featured in US magazines in the same years.[27] Garrison's examples were presented using exhaustive measurements and detailed descriptions, on the pattern of architecture treatises used to teach students classical rule and proportion: the purpose could only have been to facilitate the reproduction of these buildings—or at least some of their scenographic details— by architects in the United States.

In the nineteenth century, the methods of J. N. L. Durand initiated a change in architectural education at the École des Beaux-Arts, insisting that architectural composition should be based on the selection of elements from a compendium of styles, as if from a catalogue. By the twentieth century, images reproduced in the mass media would function in an analogous way. The manner of appropriating these images was inscribed within a modern process and implied an even more profound shift in architectural design. Architects adopted recognizable and familiar images from non-traditional sources, outside the parameters of the

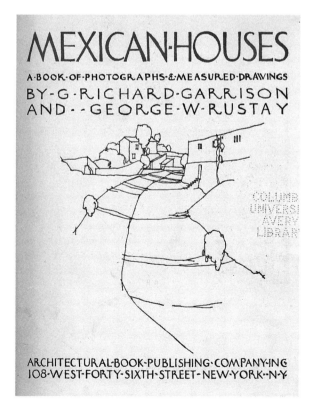

10.12 Title page of Richard Garrison's *Mexican Houses*, 1930

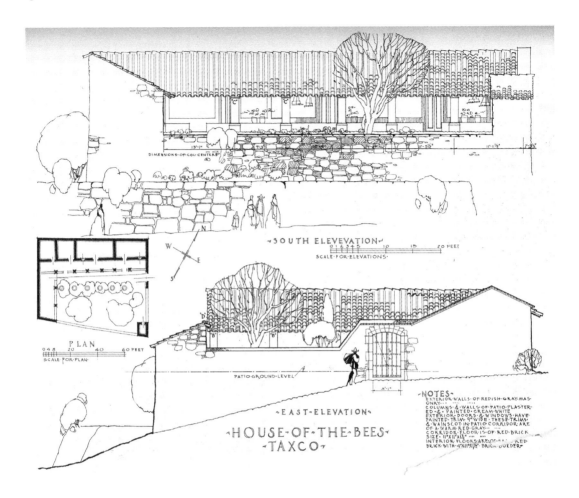

·SOUTH·ELEVEVATION·

10.13 Measured drawing of
The House of the Bees,
Colonial Taxco, in Garrison's
Mexican Houses, 1930

academic field, and used them in experimental artistic searches for a local and modern expression. The case of Spanish-colonial-inspired architectures discussed in this chapter is perhaps one of the first instances of this phenomenon.

The fact that neocolonial architecture was actually contemporary to modernism cannot be made to fit within an evolutionary conception of history. Breaking with this approach allows us to consider the possibility that certain images and ideas may have moved between these two languages, which by all appearances were the exact opposite. Many of the ideas behind the "commercial" Californian styles, and behind the image of the Mexican colonial in the United States, were—once diffused as blurry images in the mirror—eventually incorporated into modernist Mexican architecture. The earliest Functionalist expressions had echoes of these images in the work of two radical Functionalist architects: Juan O'Gorman and Juan Legarreta.

We know that the idea of recuperating Mexican traditions and popular arts through modern art was one of the imperatives of cultural projects in post-revolutionary Mexico, in the search for a new identity for the country. However, it seems likely that the way in which these traditions were recuperated was based on the images that the other, or the outside—in this case the United States—was expecting to find. The elements that caught the attention of the United States

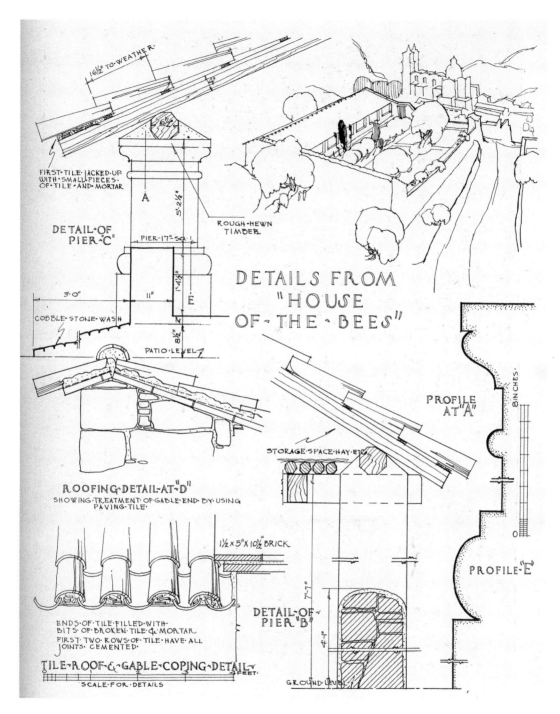

10.14 Architectural details from The House of the Bees, Colonial Taxco, in Garrison's *Mexican Houses*, 1930

10.15 Drawing of a Mission-style house by Jorge González Camarena used on the cover of *Cemento*, 1930

were those that made Mexico different from elsewhere, that distinguished it as a nation. In this way, the revaluation of colonial architecture by US architects, by the tourism industry, and by 1930s Hollywood was influential in Mexico because it gave local architects a justification, other than history, to delve into the colonial past in search of an identity, investing it with a different inflection to the official neocolonial of the 1920s. For this reason, it is no exaggeration to think that the images of this ambiguous and complex colonial that modernist or Functionalist architects saw reflected in the mirror—that is, in the mass media—influenced their work.

The view from the outside can be useful for developing a better understanding of the self. The images of the colonial moved between the United States and Mexico in both directions. Mexican architects were aware how their cultural heritage was being interpreted through the construction of the image of the US Southwest and the creation of the Californian neocolonial styles. In its dissemination in mass media, they saw themselves reflected: it perhaps was their own image seen in the mirror that they were trying to adopt and adapt. Perhaps this is always the way that identities are fabricated.

Notes

1 See Juan O'Gorman, "El Departamento Central inquisidor de la nueva arquitectura," in *La palabra de Juan O'Gorman* (Mexico City: UNAM IIE, 1983), 118.
2 See Jesus T. Acevedo, *Disertaciones de un arquitecto* (Mexico City: Mexico Moderno, 1920).
3 Jorge Alberto Manríquez, "México se quiere otra vez barroco," in *Arquitectura Neocolonial: América Latina, Caribe, Estados Unidos*, ed. Aracy Amaral (São Paulo: Fundação Memorial da América Latina, 1994), 46.
4 For more on the image of Mexico as portrayed in early Hollywood movies, see Emilio García Riera, *México visto por el cine extranjero* (Mexico City: Era, 1987).
5 Margarita de Orellana, *Filming Pancho: How Hollywood Shaped the Mexican Revolution* (London: Verso, 2009), 7.
6 The most violent struggles of the Mexican Revolution ended around 1920, but were followed by the Cristero War (1926–1929).
7 See Robert Winter, *Toward a Simpler Way of Life: The Arts & Crafts Architects of California* (Berkeley: University of California Press, 1997).
8 For more on this topic, see James Oles, *South of the Border: Mexico in the American Imagination, 1917–1947* (Washington: Smithsonian Institution, 1993).
9 While the pantry and the breakfast room were common features in Anglo-American houses, the sewing room was not, which seems to suggest that this was a kind of Mexican imagination of an Anglo-American house, perhaps stemming from the fact that the sewing machine was similarly a modern innovation from the United States.—Ed.
10 For more on this topic, see Rafael Fierro Gossman, *La gran corriente ornamental del siglo XX: una revisión de la arquitectura neocolonial en la ciudad de México* (Mexico City: Universidad Iberoamericana, 1998).
11 "Antonio Ruiz," *Forma* 1, no. 4 (1927): 40–44.
12 Ibid., 40. This and all subsequent translations are by the author.
13 "Noticias del extranjero," *El arquitecto* 1 (June 1925).
14 Juan O'Gorman, *Autobiografía* (Mexico City: Pértiga, 2007), 101.
15 Federico Sánchez Fogarty, "La farsa de Los Ángeles," *Tolteca* 21 (1931): 301.
16 Federico Sánchez Fogarty, "Publicidad institucional," *Edificación* 4, no. 2 (March–April 1937): 30.
17 See, for example, the work of Ignacio Díaz Morales in Esther Born, *The New Architecture in Mexico* (New York: The Architectural Record, W. Morrow Co., 1937), 11, 105–109.
18 José Villagrán García, "Panorama de 50 años de arquitectura mexicana contemporánea," *México en el Arte* 1, no. 10–11, (1950): 183–199.
19 Carlos Obregón Santacilia, *50 Años de arquitectura mexicana (1900–1950)* (Mexico City: Patria, 1952), 76–78.

20   Ibid., 78. "Syrian-Vaseline" is apparently a reference to a specific house, the Casa Ayub, which belonged to a prominent Lebanese family in Mexico City: see http://polancoayeryhoy.blogspot.com/2011/04/la-casa-ayub.html—Ed.

21   Israel Katzman, *La Arquitectura Contemporánea, precedentes y desarrollo* (Mexico City: Instituto Nacional de Antropología e Historia, 1964), 77–98.

22   Ignacio Diaz Infante, who designed hotels and train stations in the Spanish colonial style, was one of his teachers.

23   Álvaro Aburto, in *Pláticas de arquitectura Mexico: 1933* (Mexico City: Sociedad de Arquitectos Mexicanos, 1934), 129–133.

24   Born, *The New Architecture in Mexico*, 68–69, 91, 99–109.

25   Sylvester Baxter, *Spanish-Colonial Architecture in Mexico* (Boston: Millet, 1901); subsequently translated into Spanish as *La arquitectura hispanocolonial en México* (Mexico City, 1934).

26   G. Richard Garrison and George W. Rustay, *Mexican Houses: A Book of Photographs and Measured Drawings* (New York: Architectural Book Publishing Company, 1930). The book has recently been reissued as *Early Mexican Houses* (Stamford: Architectural Book Publishing Company, 1990).

27   "Federal Schools of Mexico," *Architectural Record* 75, no. 5 (May 1934): 444–446.

11

AVIATION, ELECTRIFICATION, AND THE NATION

Visions from Colombia and Chile[1]

Hugo Mondragón López

The magazines *Arquitectura y Construcción* (*AyC*) and *Proa* (meaning "Prow") both began publication in the mid-1940s, and their first issues demonstrate the optimistic spirit of the postwar period. *AyC* was published in Santiago, Chile, between December 1945 and August 1950, with a total of eighteen issues. *Proa* first appeared in Bogotá, Colombia, in August 1946, and in November 1951 issued a retrospective compilation after five years and fifty-three issues, thereby concluding its first phase of publication. Besides their contemporaneity, the magazines shared the determination to articulate a synthesis between economic modernization and nationalism; how the editorial teams developed opposing strategies to achieve the same goal is explored in this chapter.

A striking feature of the articles published in the first issues of *Proa* is that a good number are dedicated to the architecture of the past, specifically to Colombia's colonial architecture. Strictly speaking, according to the standard historiography these should not be appearing at this date. However, their existence was neither accidental, casual, nor random—much less contradictory to the task that the editors had undertaken: to set down the foundations of a new architecture.[2]

For Carlos Martínez, editor/publisher of *Proa*, in the period from the foundation of Colombia's earliest cities to the 1920s, the impossibility of constructing a notion of national unity based on geographical proximity was the outcome of a spatiality of dispersed urban nuclei—cities which, in the best case, were connected by rudimentary and dangerous roads (as revealed by period engravings reproduced in the magazine), or lacked them altogether. Martínez linked this dispersal to the convergence of two phenomena: the particularly uneven

11.01 "Riverine roads: The Magdalena, marvellous pathway, route of gold and of coffee." *Proa*, 1949

topography of the Colombian territory, crossed by three mountain ranges in a
north–south direction, and the fact that "[t]he Spanish settlers didn't—as the
Romans did—take the precaution of establishing roads able to be traveled by
wagons on the territories that they conquered."[3] On one hand, this uneven
topography had made Colombia an extremely rich and well-supplied territory in
environmental terms, but on the other hand, these same conditions had become
the biggest obstacle for the emergence of a "larger and more important
civilization." To provide an image of how isolated Colombian cities were from
each other, Martínez included an evocative space–time comparison: "[t]he statue
of General Santander . . . took less time to get from Paris to Cartagena than it
[then] took to get to Honda, and this was shorter than the time needed to transfer
it from there to Bogotá."[4]

Colombia was shown not only as a country excluded from the possibilities offered
by the modern world—where "the highway had been the vehicle of other big
colonies, such as Canada, the United States, Argentina, Australia"[5]—but also
occupied by isolated cities, a sort of system of solitudes that prevented the
process, and the consciousness, of converting itself into a unified nation. Various
articles published by the magazine linked the lack of a sense of a nationhood
experienced through geographical unity to the difficulties of building an efficient
road network. As Jorge Arango and Carlos Martínez noted in their 1951 compilation, *Proa* believed that it was ineffective methods of transportation and the
"painful circumstances of the trips that for over three centuries stimulated the
formation of regions that became self-sufficient as in a progressive autarchy."[6]
Their systematic reading of the articles on historical topics revealed that, for *Proa*,
the nation's particular populating strategy prior to 1920 could be synthesized in

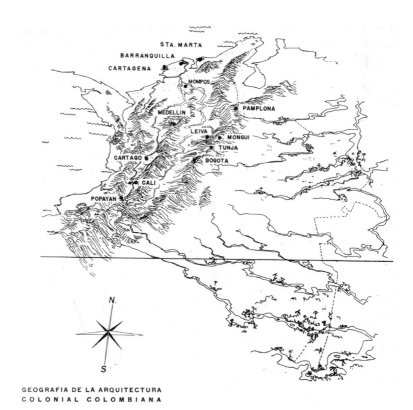

11.02 "Geography of
Colombian Colonial
Architecture": colonial-era
towns were scattered
throughout the Andes

GEOGRAFIA DE LA ARQUITECTURA
COLONIAL COLOMBIANA

two words: dispersal and isolation. The magazine saw Colombia as "a series of isolated prosperous parts whose constant progress had required painful and tenacious efforts."[7] But these conditions changed radically with a single event: the arrival of the airplane. This was at once the symbol of the arrival of modernity and the instrument that would allow Colombia to achieve national unity. As a new means of transportation, the airplane released the cities from what Arango and Martínez termed their "centuries-long seclusion." As *Proa* wrote in 1948:

11.03 Bogotá, November 1946: in twenty years, the airplane completely modified Colombia's complex topography. Photograph by Sady González

> One day Colombians . . . lifted their eyes to the sky and saw a redeeming sign. It was an airplane, that, for the first time in a morning in July 1919, leaped among the clouds of the Bogotá sky. From then on things changed: mountains which were until then unbroken obstacles, disappeared. The tense relationship between the earth and man all of a sudden became friendlier.

> The country became flat and the distances that before took weeks or months, became hours or fractions of an hour. All that was separated, vast, unreachable, became familiar. The most distant Colombians could be greeted, could get to know each other, and strengthen their commercial relationships. When America's first organized air transportation company was born—abruptly and without completing the intermediate stages—people, commerce, and agricultural production got off the mule and entered the shaking airplane cabin.[8]

For Martínez, economic modernization implied a strategic domestication of the landscape with clear productive and economic ends, including networks of highways, railroads, irrigation systems, hydraulics, and so on. These projects had always been very difficult if not impossible to execute, delaying the development of an economic modernization based on the continuous and rapid movement of people, but especially of goods and—of course—capital. According to *Proa*,[9] the first signs of a modernization of the territory appeared just as aviation consolidated itself as a means of transportation, and the enormous demographic, industrial, and commercial growth of the country between 1920 and 1950 was due to its arrival: "the airplane turned Colombia into a flat country and Bogotá into a port."[10] With its speed, the space–time relationship between Colombian cities—where a smaller distance had not always meant a shorter travel time—now became directly proportional. This new proximity brought a unity of clearly economic origin built thanks to the airplane's network of "routes," but it surpassed the purely utilitarian and productive dimension to create a social, political, and cultural unity—in other words, a nation. For Martínez, modernity and nation were paired and depended on a single group of actions on the territory that, without physically modifying the geography, transformed the way it was experienced. From 1946 to 1951, *Proa* disseminated the idea that a nation only existed within the framework constructed by economic modernization, and that there was no possible nation beyond the limits that this imposed.

With the transition from a world of dispersed and unattached urban nuclei to the self-image of a unified nation, *Proa* promoted the idea that only architecture produced after this consummation could truly be called "Colombian." For Martínez, this meant that only modern architecture built after 1945, made in Colombia, by Colombian-trained architects. Martínez explained that:

> very recently there was no need for architects [and] [w]hen, on odd occasions, it was necessary to build an important building, such as the National Capitol of Bogotá [1847–1915], foreign architects were imported and in some cases complete plans were studied in France and in Italy.[11]

Furthermore, the republican period "doesn't offer, generally, any architectural examples of importance," perhaps because "[t]here wasn't a rich and prosperous nineteenth century—unlike in Chile, Argentina, or Uruguay—in which to change the sober colonial demeanor for a more splendid one of French or Italian appearance."[12] But not all other modes of architectural production were discarded. If the buildings of the colonial period had been made without the intervention of architects, then this anonymous architecture was a collective creation and it therefore possessed the purest expression of the spirit of the Colombian people. As Martínez and Arango argued:

> The colonial architecture in Colombia is sober and austere.... Our architecture, in comparison to others of the Americas, is less restricted and tied to symbolic obstacles; it is the least eclectic, the most Creole, the least imported, thus maybe the most functional.[13]

11.04 Airline advertisements promoted the notion of the airplane as builder of a new transportation network

11.05 Thomas Reed, National Capitol, Bogotá, 1847–1915

In this way, colonial architecture became the Colombian equivalent of what classical or medieval architecture had been for modern architects in Europe who were interested in building links with tradition.

Colonial architecture presented Colombia's architectural character in its purest form; on the other hand, contemporary architecture could be interpreted as a natural evolution of certain archaic principles already found in colonial buildings, as rediscovered by Martínez.[14] This supra-historic continuity allowed Martínez to select from the multiple experiments of contemporary architecture those that did not contradict the character and the values of colonial architecture, thus encouraging architects to experiment further in this direction. This will to establish a bond between colonial and contemporary architecture appears in *Proa*'s description of a rural chapel:

> In the feelings of many modern architects, the functional aspect in construction relies on the logical use of a region's materials. This nice rural chapel with its modesty is an admirable example set down to those parish priests that prefer the pomp and the fanciful over the purely logical and rational.[15]

11.06 (below left) A street in Villa de Leiva, and interior of a residence, Tunja, published in *La arquitectura en Colombia*, 1951

11.07 (below right) Convento Mongui, and Hospital Tunja, published in *La arquitectura en Colombia*, 1951

In Colombia, the possibility of building a connection to the Pre-Hispanic past was not very productive because the indigenous people of Colombian territory did not build any monumental constructions that lasted over time.[16] According to Martínez, the austere character of Colombian colonial architecture was somehow determined by this absence of a rival for the Spanish conquerors to compete with, as had been the case in Mexico and Peru. Nonetheless, he did not fail to recognize that a good part of the constructive methods used during the colonial period were of indigenous invention: "This was, in general terms, the architectural legacy of the Indians. It was a simple contribution, without valuable examples of plastic order or of any plan composition."[17]

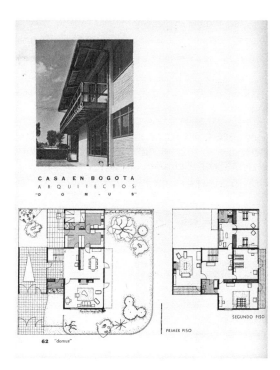

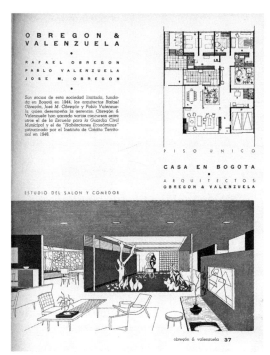

Martínez referred to the peak of the nationalist styles of the 1930s as the event that had helped to "break the umbilical cord that sentimentally united Colombians with the colonial house," thus preparing the ground for a new architecture, whose arrival was sudden and without any intermediate stages: as the mule's back was replaced by the aircraft cockpit, this new architecture did not replace "stylish [architecture], as in most countries, but the uncomfortable colonial house."[18] Proa's historical articles were like the mesh of a sieve, separating and selecting fragments that were desirable. The mesh caught the values of colonial architecture which could be re-utilized: rationality, sobriety, austerity, functionality, simplicity, and the economy of financial and expressive means. These words indistinctly described the character of colonial as well as contemporary architecture, and were useful for Martínez to express his non-temporal conception of the modern in architecture. The confrontation between contemporary and colonial architecture appeared dramatically in the compilation published by Proa in 1951. The book's subtitle proposed a deliberate historical leap juxtaposing these architectures, by eliminating all architectural production from 1810 to 1945 in a single blow, thereby establishing a radical historical continuity between colonial and contemporary architecture. As the introduction noted: "350 years of the colonial era and five years from 1946 to 1951 correspond to the most important periods in Colombian architecture."[19] One notable example of this crossed influence as synthesizing the Colombian modern was Casa en Bogotá by Domus Architects. With its striking structure of concrete vaults, this house had caught the attention of Le Corbusier, who carefully sketched them on his trip to Bogotá, but it was the long wooden balcony which Martínez discussed in an exhaustive commentary, translating this typical colonial element into a modern language. Proa also published numerous examples of houses that employed a plan with a central courtyard as a hierarchical element, which Martínez similarly considered part of the colonial Colombian tradition transplanted to modern architecture. One example was the home for the architect José María Obregón, published in June 1949. With these and other examples it becomes clear that Proa was the work of

11.08 (above left) "Domus" Architects, house in Bogotá, 1951: modern house "quoting" the long balcony typical of colonial buildings

11.09 (above right) Obregón and Valenzuela, residence in Bogotá for Alvaro López, 1950: modern house "quoting" the patio typical of colonial buildings

a modernist-nationalist—Carlos Martínez—who saw both these dimensions as complementary and necessary to the configuration of a program of architectural modernity in Colombia.

In Chile, the relationship between economic modernization and geography was perceived quite differently. Very early on, Chile was able to build an efficient network of roads and railways throughout its long and narrow territory. In contrast to Colombia, Chile's problems of territorial organization were not concerned with the lack of transportation networks, but with the excessive centralization of economic and political activities in the Valparaíso–Santiago axis. As the editors of *Arquitectura y Construcción* informed their readers in 1946, 70 percent of the country's population and industries were concentrated in these two cities, with a further 10 percent in Concepción, and the remainder throughout the rest of the country:

> Is this excessive concentration convenient? Was it initiated for industrial reasons or, as seems more probable, is it a simple consequence of the political, economic, and other centralisms which the country has suffered? Doesn't this produce a disproportionate growth of the capital and its port to the detriment of the rest of the country?[20]

The editors supported decentralization, but how could this be applied? The answer appeared in the context of developing energy sources for industry.

Thus *AyC* published several articles on industrialization by authors with a close relationship to the government's Corporation for the Development of Production (CORFO). These presented industrialization as a means to progress beyond the production of raw materials for export—perceived as maintaining Chile in a state of economic colonialism—and toward the production of manufactured goods. As Sergio Vergara highlighted in an article titled "Chile: Country with an Industrial Destiny:"

> People that are not able to transform the production of raw material into manufactured goods are only partially free. . . . How heavily does this economic semi-servitude weigh on most South American countries that are positioned by their capacity as simple extractors of raw materials![21]

Vergara asserted that for Chile industrialization was not merely an option but the only road that it should follow. This was symptomatic of the moment: the political and cultural project of constructing a modern nation depended on industrialization.

Vergara pointed out that the development of new energy sources was funda- mental for the construction of an industrialized nation—not just coal and petroleum, but most particularly electricity, which was regarded by *AyC* as the most suitable and efficient form of energy. In this context, he highlighted CORFO's National Electrification Plan, which had considerably increased the country's electricity production. As *AyC* affirmed, this was essential for moving from an agrarian and mining economy to an industrial economy. In an article dedicated to promoting the work of the recently created National Electricity Company (ENDESA), *AyC* argued that the core of CORFO's politics "can be expressed in the statement that a country's industrial production capacity is in practice independent from the population and . . . depends exclusively on the avail- ability of the country's mechanical energy."[22] In *AyC* this was particularly linked to hydroelectric power stations, which were a significant element of the program for industrial decentralization it imagined. In the article on ENDESA,

11.10 Cover of *Arquitectura y Construcción*, 1949

CHILE: PAIS DE DESTINO INDUSTRIAL

Artículo escrito especialmente para nuestra revista por don Sergio Vergara Vergara, autor del libro "Decadencia o Recuperación — Chile en la Encrucijada", de reciente publicación.

25

11.11 "Chile: Country with an Industrial Destiny," *Arquitectura y Construcción*, 1946

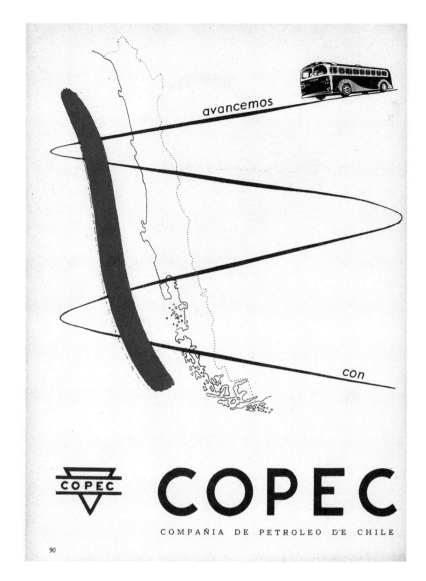

avancemos

con

COPEC

COMPAÑIA DE PETROLEO DE CHILE

11.12 "Let's move forward with COPEC": navigating Chile's elongated geography with the assistance of the Chilean Petroleum Company. *Arquitectura y Construcción*, 1946

Chile—articulated in an economic–industrial sense—was bounded by the strip of land between Coquimbo-La Serena to the north and Puerto Montt to the south, leaving the northern desert and Patagonia outside of the national territory. The central section of this mutilated territory was divided into six regions defined by access to water resources, each with a hydroelectric power station as a nerve center. (Three were already completed, and three projected.) The national territory was conceived as an intricate network of transmission cables with focal points in these power stations, each of which provided regional coverage and also communicated with each another, building a continuous nervous tissue across the selected territory. In this way, the articulation of the national economy's project of industrialization gave birth to a new cartography that was strongly defined by the absence of a single center—which, under the old system, was Santiago-Valparaíso—in favor of a polycentric territorial system.

By the time that *AyC* appeared in 1945, the future, progress, technological advancement, and modernity were embodied in the project of industrialization

11.13 Santiago refrigerating
plant, for the Agricultural
Export Board, *Arquitectura
y Construcción*, 1946

which Chile had promoted beginning in the 1930s. With the creation of CORFO in
1939, a new phase of this process had begun, characterized by proposals for
national development. The pages of *AyC* picked up these themes with articles on
an important group of industrial buildings, the development of the construction
industry, and an enormous quantity of advertising announcing the development
of a national industry producing construction materials. From the second half
of the nineteenth century and until the celebrations of the Centennial of Independence in 1910, metal structures were considered the material signs of progress
in construction technology; by the 1930s, these had already been relieved from
that function by structures in reinforced concrete.

The consolidation of this shift can be verified with particular precision through
analysis of the advertising published in the magazine over its lifetime, between
1945 and 1950. This indicates the centrality of the production of cement—and to a
lesser extent the production of iron and glass—in the construction industry's self-
image. Through these advertisements, the Chilean construction industry tried to
grant a social use to construction materials: reinforced concrete was championed
as a material embodying modern construction, with its natural destiny to build

11.14 Central aviation hangars, *Arquitectura y Construcción* 1946

11.15 Advertisement for Lamifun iron suppliers, *Arquitectura y Construcción*, 1947

bridges, industrial warehouses, and "skyscrapers." In the pages of *AyC*, these reinforced concrete structures were presented not only as a significant indicator of Chile's material progress, but also as the consolidation of the nation's commitment to the project of industrialization.[23] As a consequence, the aesthetic valuation of reinforced concrete structures promoted by *AyC* magazine did not face any resistance from the architectural profession or the culture more broadly. During the period of the magazine's circulation, reinforced concrete's mathematical, precise, and plastic beauty, and the value of its genuine contribution to national development, became the prevailing discourse within the discipline. In the issue of March 1946, devoted to the publication of industrial facilities, several images showed exposed reinforced concrete structures, one example being the Electrical Materials Factory ELECTROMAT. Moreover, the use of reinforced concrete with an aesthetic or plastic purpose was probably first applied in three emblematic works of the 1940s: the Chillán Cathedral, a large central nave built with parabolic arches of reinforced concrete, and two workers' clubs—the Hogar de Defensa de la Raza and the Hogar Hipódromo de Chile—which were published in great detail in *AyC*.[24] Advertisements for the Juan Soldado Cement Industry in La Serena published in October 1946, or photos of the construction of the ENDESA dams in March 1946, among others, indicate the importance given by the editors to a modernization process that aimed to develop a larger, decentralized territory.

The two magazines are united by their intention to articulate a connection between economic modernization and a national political and cultural agenda. In the postwar period, many Latin American countries felt that their economies, now industrialized, were stronger than in the interwar period. In ideological terms, this brought into focus a paradoxical synthesis between two opposing conceptions. On the one hand, economic modernization linked to progress, and to the future; on the other hand, a nationalism connected to tradition, to heritage, to culture, and to the past. These two magazines' shared purpose was to articulate a program of nationalist modernization; nonetheless, their two contrasting modes of expressing this project were emblematic of the two ideological poles between which the architectural culture of the region oscillated. In the case of *Proa*, the program of national modernization led toward a recovery of the architecture of the past, demonstrating how the character of modern architecture shared a common set of values with colonial architecture. In the case of *AyC*, national modernization chose the future as its battlefield. Modern reinforced concrete structures were spread throughout the pages of the magazine as monuments evoking a future nation that cohered around the project of economic modernization. Furthermore, while Carlos Martínez was interested in showing how the development of commercial aviation enabled various Colombian cities to feel part of the nation for the first time, the editors of *AyC* were more concerned with championing territorial decentralization for purely economic ends.

In this sense *Proa* and *Arquitectura y Construcción* would come to represent two emblematic positions characterizing the development of modern architecture in Latin America during the twentieth century. On the one hand, balancing the modernizing impulse and tradition, while on the other, dissolving culture in the modernizing project. In the first case, the nation was a cultural project, of the past and of permanence; in the other, the nation was a development project, of the future and of change.

Notes

Translated from the Spanish by Gisela Frick.

1   This chapter is based on my earlier research, including: Hugo Mondragón, "Chilean Architectural Culture and its Periodicals, 1930–1960" (Fondecyt Project no.1990449,

2010); "Architecture in Colombia 1946–1951: A Critical Reading of *Proa Magazine*" (MArch thesis, Pontificia Universidad Católica de Chile, 2002); "The Speech of Modern Architecture, Chile 1930–1950: A Construction from Periodicals" (PhD diss., Pontificia Universidad Católica de Chile, 2010); Hugo Mondragón and Andrés Téllez, "*Arquitectura y Construcción*, Chile 1945–1950: A Magazine of Modern Architecture."

2   Architects of the time declared themselves openly and resolutely against historical architecture, especially of the recent past, while, according to Silvia Arango, they had a reluctant respect for colonial architecture: "Out of this tabula rasa of the past only colonial architecture emerged clean, and only because it shared some characteristics with the features they defended." Silvia Arango, *"La evolución del pensamiento arquitectónico en Colombia, 1934–1984."* (13ro Anuario de la Arquitectura en Colombia, Sociedad Colombiana de Arquitectos, Bogotá, 1984).

3   "Los caminos de Colombia," *Proa* 22 (April 1949).

4   The magazine's almost obsessive concern with this topic finds a probable explanation in its belief that: "Transportation routes determine the fate of cities. A great city does not exist without large transportation networks. Ground, marine, or air routes help to structure the physiognomy of the urban nuclei. If we want to inform ourselves about the origin and development of a city it is necessary to observe its location and its geographic horizon, always tied with its transportation routes." Ibid.

5   Ibid.

6   Jorge Arango and Carlos Martínez, *Arquitectura en Colombia: arquitectura colonial 1538–1810, arquitectura contemporánea en cinco años 1946–1951* (Bogotá: Ediciones Proa, 1951), 9.

7   Ibid.

8   Ibid.

9   This was *Proa*'s version of growth. "Today [1951] with transportation three hundred times faster, the confused situation of isolation has disappeared. In three decades, the most important urban centers in the country have increased the volume of their industries and trade a hundredfold. Fiscal profits have increased 20, 50, or 100 times over . . . this combined process, whose effects are evident, has resulted—due to the great number of urban nuclei—in Colombia's epithet of the country of cities." Arango and Martínez, *Arquitectura en Colombia*, 9.

10   *Proa* 22 (April 1949).

11   Ibid.

12   Ibid.

13   Ibid., 18.

14   The word "evolution" appears in the titles of two of the articles examined in this section: "Evolution of Styles in Colombia" and "Study of the Evolution of Bogotá's Streets."

15   *Proa* 4 (January 1947).

16   Nevertheless, during the 1930s the artistic group Los Bachués had experimented, without much success, with the metaphor of *miscegenation*—the incorporation of indigenous motifs into contemporary buildings. The vindication of colonial architecture as the earliest form of Colombian architecture made by Carlos Martínez in the mid-1940s is partly explained by the failure of such experiments.

17   Arango and Martínez, *Arquitectura en Colombia*, 14.

18   Ibid., 32.

19   Ibid., 33.

20   "Generalidades Estadísticas de la Industria Chilena," *Arquitectura y Construcción* 4 (March 1946).

21   Sergio Vergara Vergara, "Chile: país de destino industrial," *Arquitectura y Construcción* 4 (March 1946).

22   Ibid.

23   Architectural structures in reinforced concrete had been admitted into the discipline and culture at least since the centennial celebrations in 1910, but still under the garb of historical styles. Belonging to this period are works such as the Museum of Fine Arts, the National Library, or the main building of the Universidad Católica, all works of Beaux-Arts expression with supporting structures built of reinforced concrete. By the late 1920s, the first reinforced concrete bridge was built over the Mapocho River, the main waterway that passes through the city of Santiago, and immediately became

the new sign of constructive progress in Chile. The inauguration of the Puente del Arzobizpo in 1929 not only overcame the old obsolete metal bridges that were built on the same river, but its entry into service was an event celebrated in the pages of magazines—both popular and specialized—as marking the beginning of a new era of material progress for the city and the country.

24 While the Chillán Cathedral used concrete for the expressive power of its monumental structure, in the workers' clubs a more domesticated, or at least less dramatic, use of the material is observed. These are works that exhibit an almost academic, careful plastic use, trying to approach some of the artistic expression of Le Corbusier's works derived from the use of the Dom-Ino prototype.

12

MARIO PANI'S HOSPITALITY

Latin America through *Arquitectura/México*

George F. Flaherty

Architect Mario Pani was unhurried in embracing the work of his Latin American colleagues in the journal he founded and edited, *Arquitectura*, which would become the longest running (1938–1978) and, arguably, the most influential architecture journal in Mexico.[1] That, as Latin Americans, they were colleagues at all was still crystallizing for Pani—starting in the late 1930s—as was the practice of an expansive form of architecture. This expansion was both territorial, looking at not just buildings but entire cities and beyond, and jurisdictional, claiming professional domains of technical and social expertise. Mexico's modernist architects continued to grapple with the notion of a national school of architecture while at the same time reproductions of their work traveled through increasingly permeable—though never wide open—borders. Pani's treatment of Latin American modern architecture in *Arquitectura* was a speculative enterprise, depending on a rotating network of correspondents as well as secondhand reports and reproductions rather than established bureaus and beats. The journal played a major role in eventually consolidating a modern Mexican architecture in connection and contrast to Latin American examples; or just as often, with no affinity at all. Pani's uneven approach does not appear extraordinary when held up to the wavering light of Latin Americanism. Historically, interest in hemispheric solidarity has rarely been disinterested. *Arquitectura* did not fulfill a Bolivarian fantasy. That is not to say, however, that there was not much to gain and lose in how Latin America—and Mexico—were defined in its pages.

The journal's interest(edness) in Latin America coincided roughly with a period known as *apertura* (aperture, or opening) in Mexico at mid-century. Beginning in the 1940s and intensifying as the 1968 Olympic Games in Mexico City approached, Mexican elites looked outward more publicly than during the first decades after the revolution (1910–1920). The state also revised its macroeconomic policies, shifting from an import-substitution industrialization model, which was showing signs of distress, to what it called "stabilizing development," expanding the banking sector to attract increased capital investment, especially from abroad.[2] Much of this money flowed toward urban and transport infrastructure, leading to a real estate boom, especially in Mexico City, which was also experiencing unprecedented population growth. This period came to be known as the "Mexican Miracle," although it was more a leap of faith than a comfortable reality for most Mexicans. Given the journal's long run and many contributors, a comprehensive analysis is beyond the scope of this chapter.[3] The charge instead is to identify its overarching territorial and jurisdictional logic while respecting its heteroglossia. Pani, as founding editor but also urban theorist and politically connected real estate developer, will serve as the organizing figure, standing for the relationship between material and discursive interventions required to constitute a Latin American modern architecture—if such a category ever existed.[4]

Pani created his own *apertura*, which gestures to passage as well as restriction, within *Arquitectura*. In 1949, the journal changed its name to *Arquitectura/México*, at once a response to competition from other, similarly named journals in Latin America, but also marking the tension between an increasingly cosmopolitan discipline and the nation-state. Well aware that architecture was consumed globally, whether in person or media, Pani positioned himself as translator and host to Latin American architecture. This was a "conditional hospitality," however, to borrow from Jacques Derrida, accommodating new affinities while maintaining long-held Mexican exceptionality. As Derrida's philosophical investigation into the ethics and politics of hospitality suggests, the interaction of host (Pani) and guest (Latin American architecture) is mutually constitutive though not necessarily symmetrical or stable.[5] This was based as much on hemispheric feeling as a calculation of finite resources and competing interests. In the first issue, Pani wrote that the journal, "goes out to the Spanish-speaking world and particularly Mexico, hoping for a warm welcome." Pani would figuratively stand at the threshold to see what reception the journal (and Mexican architecture) received—as well as to determine what from abroad would enter its pages. The role of host guaranteed certain responsibilities and rights.

Pani left Mexico as a young boy to follow his father, who worked in the Mexican diplomatic corps, serving in Antwerp and Geneva among other cities. He traveled extensively in Europe and learned French, receiving what he later called an "internationalist" education.[6] Trained at the École des Beaux-Arts in Paris in the 1920s, Pani directly observed the initial currents of Europe's modernist architectural culture and developed his practice in response. In the journals that Le Corbusier and André Bloc created, *L'Esprit nouveau* (1920–1925) and *L'Architecture d'aujourd'hui* (1930–present) respectively, he found imperfect models. While he recognized the power of mass media to shape professional and public perception, he consciously distanced himself from what he saw to be the ideological partisanship of such avant-garde journals. Pani wrote in the first issue of *Arquitectura* that he was "shedding all exclusive doctrine, [and] sectarianism," replacing it with ostensibly impartial "documentation" of developments abroad.[7] With youthful enthusiasm as well as cosmopolitan privilege, Pani saw the world getting smaller and more similar through new communication and construction technologies. He wrote, "One lives in the same way, with the same prejudices and same requirements, in Cape Town and in London, in Mexico City and Shanghai."[8] Architecture, he added, was not immune: "it internationalizes itself." Pani would play a significant role in this process which looked forward and into the past, although, tellingly, he did not at this time choose to include any other Latin American coordinates beside his home city. Early issues understood the international as exclusively Western European, with the first issue offering a modern-day Grand Tour: Rome to Paris to London to Stockholm, back to Rome, and on to Paris again. In the third issue, Pani invited his mentor from the École, Georges Gromort, to defend baroque architecture, and in the fifth he was invited back, this time to conclude, "Palladio was right." Of course, this history was marshaled in service of the Functionalism that Pani and others advocated for post-revolutionary Mexico. Vladimir Kaspé, a close friend from the École still in Paris, also conducted interviews with the newer "masters," including Le Corbusier and Auguste Perret.[9]

Pani would eventually decide that the world was not so homogeneous after all, although this did not automatically translate into a hemispheric or global orientation. Initially relegated to a few pages toward the back of the journal, Europe was progressively provincialized in favor of Mexico. In 1944, he began to produce entire issues devoted to contemporary national architecture. *Arquitectura* shaped and was shaped by Mexico's emerging architectural culture, which included the creation of professional organizations, training curricula, and think tanks, in which Pani often played a founding role. This professionalization was informed by Pani's interest in urban planning as a supplement to architecture. Pani understood Mexico City as a networked system rather than hermetic mass, leading to such projects as his master plan for "Satellite City," a self-sufficient bedroom community built in the 1950s just north of the capital that would free up its historic core for higher value commercial and touristic "regeneration."[10] Pani's planning perspective perceived finite yet interconnected scales of intervention (local, national, international) and the political, economic, and rhetorical expediency of such identifications and crossings among them. Reflecting on those early issues on the occasion of *Arquitectura*'s twentieth-fifth anniversary, Pani took an almost contrite stance toward his early Eurocentrism.[11] This and various other anniversary remembrances transformed what was a halting process of articulating a Mexican modern architecture based on a series of professional gambles and utopian speculations into a sweeping Bildungsroman: Mexican architects slowly but surely developed their technical and aesthetic confidence— and *Arquitectura* jointly developed its editorial voice—so as to wean themselves off European models. Conversely, modern architecture from Latin America was not a source or influence to be overcome so much as a competitor on the world stage as the mostly Euro-American critics who defined the "International Style" turned their attention, temporarily, to the region.[12] Latin American architecture became an export business for a time; its consumption outside the region depending as much on humanist and capitalist claims of universality as readily identifiable differences. These two impulses neither negated one another nor required reconciliation; the contradiction ensured that Latin American (or any other "Non-Western") architecture could enter world architecture history without overtaking the category, even as virtuosic designs during the world wars suggested otherwise. Over the years, "originality" was defined in the pages of *Arquitectura* by various authors in terms of formal novelty and also embedded in a particular point of origin. Pani's *Arquitectura* strategically situated Mexican architecture within this emerging, contradictory field.

Arquitectura first addressed Latin America—if not exactly Latin America as a category of architecture—with an article on Cuba in 1941, three years after its launch. The piece was more or less inauspicious, except that it made plain architects' aspirations toward a broader social purchase. M. A. Hernández Rogers, a Cuban architect, reported on professional matters on the island that would be of interest to his Mexican counterparts, who were also consolidating their discipline around certain professional standards and in close alignment with a pro-development state.[13] Not until the journal turned its attention to South America two years later did the relationship between Mexican architects' territorial and jurisdictional expansion come into greater focus. Manuel Chacón, frequent contributor to *Arquitectura*, reported from Lima. Although he was identified on the journal's masthead as its "representative in South America," Chacón was less a formal correspondent than a colleague of Pani's making an extended trip through the area. With Europe and the United States still embroiled in the Second World War, he noted, "our pacific Republics [*sic*] are a refuge of tranquility; of work and *even culture*."[14] Deploying a lofty yet self-deprecating tone, Chacón articulated the malleable brand of Latin Americanism that would find resonance throughout the journal's forty-year span. Recognizing Lima and Mexico City as postcolonial capitals, Chacón was eager to perceive affinities between what he saw as Peruvians' struggle to define a national architecture and the debates he participated in at home. Although Chacón's article announced a shift in the journal's editorial strategy—that foreign developments would no longer be covered from Mexico but in situ—*Arquitectura* frequently conflated correspondence with commensurability. Chacón's seemingly incongruous focus on colonial architecture, at the expense of reporting on more recent construction, suggests that his concerns lay closer to home. In Mexico, the colonial was alternately a

12.01 (right) Map of
Arquitectura's projected
activities in South America
for 1943

12.02 (below) Manuel
Chacón's study of Lima's
City Hall

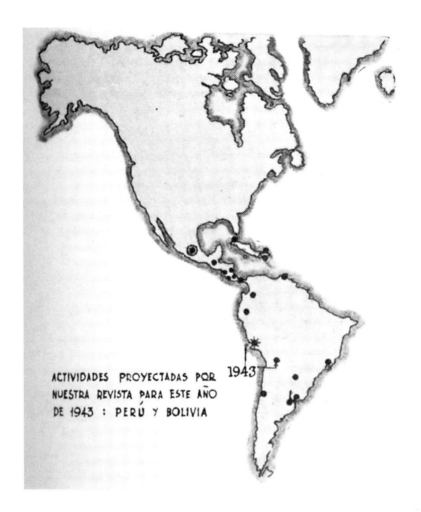

ACTIVIDADES PROYECTADAS POR 1943
NUESTRA REVISTA PARA ESTE AÑO
DE 1943 : PERÚ Y BOLIVIA

Palacio de la Municipalidad, en vías de terminación. Caso semejante al de la Ciudad de México: comprenderá también dos edificios dejando entre ellos una nueva avenida. Para no romper la escala de la Plaza de Armas, uno de cuyos lados será totalmente ocupado por estas edificaciones, asegúrase que no se construirá la torre del edificio principal. Arq. Alvarez Calderón.

sign of hybrid national modernity and bourgeois decadence, depending on whether it was the product of historic preservation or middle-class revival. Under either banner, it could be jettisoned entirely to address more pressing social issues, at the discretion of the architect–planner, of course. Suitably, Chacón's contributions appeared under the title "Arches," suggesting spans between two points—necessarily originating in Mexico.

In Corbusian fashion, Chacón had been struck by his airplane's approach into the cities he visited, first Panama City, then Cali, and finally Lima.[15] Such a perspective offered the illusion of taking a city in as a whole and perceiving its patterns "in two dimensions." Chacón imagined his airplane as a "small hotel," whose impeccable service allowed him to "surrender completely" and admire Lima through the clouds. It was a touristic rather than an aerial perspective, however, which led him to conclude that the Mexican influence was near-predominant in this "hospitable city" due to its consumption of Mexican cinema—the country's most popular export at this time. Through it "our architecture, our music, our customs, principal vices and virtues are here [their] daily bread," he wrote. Conversely, Chacón cautioned that Mexicans knew very little about Peru. As would become a trend, *Arquitectura*'s correspondents and editors implicitly and explicitly deployed Mexico as the mediating term in order to define foreign architecture, in spite of what they observed directly before them. This speculative basis was also evident in the journal's operation: from the first issue the masthead included information on newsstand and subscription pricing in US dollars (the currency of cross-border exchange), before it could possibly have had substantial international distribution.[16]

Rather than existing as a coherent category, Latin American architecture was articulated as an inconsistent series of countries of interest to Pani and his associates, relying on their personal networks and travel itineraries. This was a function not only of the periodical's financial limitations but also its liminal stance overall. Developments in Latin American countries could also serve to "objectively" justify or confirm the course of Mexico's modernist architectural culture. Brazil figured prominently in *Arquitectura* in the 1950s because, in addition to being the darling of the international architecture press at the time, Pani visited the country personally. A 1946 article on a new medical school in Santo Domingo was published as Pani, along with others, developed plans for a new campus for the Mexican national university.[17] This article is also notable because it was the first to treat a building in Latin America like its Mexican and European counterparts, as a photoessay. It opens with a half page black-and-white photograph taken from an airplane, presenting an uninhabited cluster of white, rectangular prisms floating on *pilotis* against a lush tropical landscape. A subsequent grid of smaller photographs foregrounds the play of light and shadow through covered corridors and louvered windows. Conversely, Chacón's promises of future coverage of Argentine and Bolivian architecture, which did not immediately resonate with perceivable local or personal concerns, were never realized. Antonio Acevedo Escobedo, who would later take up the editorship, summed up *Arquitectura*'s relationship to the wider world: it was Mexican architecture's "liaison" and "diplomatic service *ad honorem*," he wrote.[18] As Acevedo Escobedo's choice of words indicates, this relationship required artful forms of engagement.

The eighth meeting of the Pan-American Congress of Architects, which took place in Mexico City in 1952, offers a succinct (but by no means singular) case study in how the category of Latin American architecture was deployed strategically and could accommodate a high degree of inconsistency without negation or reconciliation. Close to two thousand architects from across the

12.03 José A. Caro Alvarez, Medical School in Santo Domingo, Dominican Republic

Americas as well as Europe attended the Congress. Its stated aim was to develop a hemispheric doctrine that, while recognizing the region's ties to "universal" architecture, addressed its specific social challenges concretely.[19] While gatherings like these were opportunities for architects to come together in person under the banner of international fraternity, so much in vogue after the Second World War, they remained organized around national delegations and competitions which rendered such identifications as the dominant frame of reference. Appropriately then, a poster for the Congress used the Pan-American Highway as a visual motif, with the road running like a spine from Prudhoe Bay, Alaska, to Tierra del Fuego. But like the actual highway, Latin America was piecemeal, subject to local conditions and by no means a direct path to "development" that such a project promised.

Carlos Lazo, who headed Mexico's two major professional architecture associations, served as Congress host, and was also a key proponent of *integración plástica* (plastic integration). This movement sought to synthesize art and architecture at a time when government patronage was shifting from self-contained projects toward urban redevelopment schemes anchored by large-scale public works. Pani's Teacher's College in Mexico City (1945) is an early example that points to the potential awkwardness of this union. Employing a Functionalist design vocabulary, it also included exterior murals by José Clemente Orozco, reliefs by Luis Ortiz Monasterio, and a colossal head sculpture from the ancient Olmec civilization with no direct historical relationship except the all-inclusive edifice of national pedagogy. More than a move to "Mexican-ize" an architecture perceived to be European in origin through the use of locally produced art (and construction materials), *integración plástica* was a discourse and practice for managing the varying scales—finite yet interconnected—Mexican architects saw themselves presiding over. Pani, along with others, advocated the adoption of urban planning techniques as a means of addressing social issues associated with rapid urbanization in Mexico, and situated the architect–planner as leader of the multi-disciplinary team needed to tackle such issues, creating a virtual home for themselves as the architectural market moved toward uncertain consolidation.

The Pan-American Congress included the regular slate of panels, tours, and exhibitions, but guests were also treated to a preview of the Ciudad Universitaria (CU), the new campus for Mexico's national university, where most of the events were held. Pani and Lazo, along with Enrique del Moral, had served as lead

12.04 Announcement for the VIII Pan-American Congress of Architects, 1952

architects to the CU, overseeing 140 architects working in small teams. *Arquitectura* published an issue devoted to the project to coincide with the Congress, providing near exhaustive photography and as well as reproductions of its planning documents.[20] Covered extensively in the domestic and foreign press, the campus came to symbolize, even before it was completed, a harmonious nexus of multiple collaborators and of modernist architects and an enlightened state, Mexican architecture's biggest patron. This did not go unnoticed even by foreign commentators. Richard Neutra, the US-based architect who would travel to Mexico regularly beginning in the late 1930s, surmised: "The epoch of the prima donna is, perhaps forever, gone. If architects want to accomplish the mission they have claimed in our time they should do as the Mexican architects: work in teams and understand that a common mission can only be realized collectively."[21] This collectivity was not, importantly, democratic or universal, and depended on the architect–planners and patrons who sought to define and deploy it at will.

Brazil Building

While Brazil did not necessarily receive the most coverage in *Arquitectura* by page count, it was closely linked to Pani's professional and rhetorical preoccupations. Beginning in the 1940s, a handful of Brazilian architects beguiled critics around the world with a brand of modernism that appeared to effortlessly combine stark and sensuous forms. Although Lucio Costa's *Plano Piloto* for Brasília did not yet exist, he and Oscar Niemeyer were already receiving considerable international media attention, fed by exhibitions like the New York Museum of Modern Art's *Brazil Builds* (1943). Among the admirers was Pani, who worked on integrating curvilinear silhouettes into his mainly orthogonal designs throughout his early career. Pani had the opportunity to examine Brazilian modernism firsthand at the end of 1951 when he was invited to serve as part of a panel of architecture judges for the inaugural São Paulo Bienal, joining several prominent colleagues, including Sigfried Giedion and Junzo Sakakura. Pani also travelled to Río de Janeiro, taking photographs of prize-winning buildings in both cities that he was eager to share with his readership in Mexico, the first mention of Brazil in the journal. Documenting the interior of the Duchen factory by Niemeyer (with Hélio Uchôa) still under construction, he emphasized its arched and coffered reinforced-concrete shell with his camera and also the linearity of its long production hall, creating a similar tension for exterior shots of Affonso Eduardo Reidy's Pedregulho housing complex.[22] After his return, Pani commented enthusiastically in *Arquitectura*, although not without some qualification: "Among the countries of the New World that do not speak English, Brazil occupies an outstanding category as a nation of drive and creative impetus . . . [and] inexhaustible originality."[23] As Chacón had done for Peru, Pani was building "arches" to Brazil.

In 1958, at the height of the architecture world's romance with Brazil, *Arquitectura* published a special issue devoted to the most recent developments in the country.[24] The issue opened with a letter from the Brazilian ambassador to Mexico—reproduced complete with signature and official seal—noting that a journal in Brazil had recently devoted an issue to Mexican architecture. This diplomatic reciprocity was meant to confirm that both countries had successfully positioned themselves as donors to Latin American and also global architectural culture. As Acevedo Escobedo noted, *Arquitectura* "pushed Mexico with Brazil as one of the profoundly original and creative countries in the Americas."[25] This shared vanguardism was a function of its maturation narrative that at once aspired to autonomy and was dependent on regional and international comparison. Neither fully integrated with—nor entirely foreign to—the idea of Latin America, Brazil, with its vast territory and natural resources, was a colossus that required

12.05 Cover of
Arquitectura's 1958
issue on Brazil

exposition. Like many of the journal's profiles, a foreign subject was best presented through a positivist recounting of its history, ancient through contemporary. This responsibility fell to Brazilian journalist–statesman Lindolfo Collor, whose article was followed by a substantive essay on the new capital city (which would be inaugurated a year and a half later) by Néstor Dos Santos Lima, another Brazilian author. Echoing *Arquitectura*'s strategic Latin Americanism, Dos Santos Lima situated Brazilian architecture not just in the Latin American but the "universal" imagination.[26] At the same time, he maintained its essential difference, which produced "maximal benefits" for Brazil and its Latin American "sister cultures." Like Pani, Dos Santos Lima recognized that to operate successfully in a globalizing architectural market, even if only through representation rather than commissions, the national architectures so tenuously constructed during the first half of the twentieth century would, going forward, want to carefully balance international appeal with internationally recognized native specificity.

Pani prided himself on translating various foreign developments that he saw as relevant solutions to Mexico City's capitalist urbanization, often changing their scale to give them a home there. These included, in his estimation, "the first truly modern hotel of its age. . . . Also the first condominium, important element for the growth of the city because it permitted higher population density and [land] utilization, with the same urban resources."[27] While his first job in Mexico was to take over design duties for Carlos Obregón Santacilia's Hotel Reforma (1936), owned by his uncle Alberto Pani, a major post-revolutionary political figure, Pani established his reputation on the basis of another type of hospitality: mass housing for the country's working poor and emergent middle class, which he called *multifamiliares*.[28] Pani possessed a competitive spirit and astutely realized that large-scale, government-patronized architecture (that also tapped into foreign financing), which he championed, was turning toward industrialization (that is, prefabrication, modularity). Indeed, his first *multifamiliar*, Presidente Alemán (1948), led to the formation of a private company that could take on complex seismic and structural engineering studies and construction. Ingenieros Civiles Asociados would become a driving force in reshaping Mexico City, including strongly advocating for a subway system that it profited from.

Brazil offered an internationally recognized model to confirm Pani's professional and social speculations.[29] He drew on its example to promote condominiums as a solution to Mexico City's overpopulation and sprawl. Such a property regime, with its vertical, multi-family organization, was foreign to the Mexican middle class as well as its legal system. Pani made his case beginning in 1953 by suggesting that Mexico was falling behind Brazil, as well as Argentina and the United States, in this key category of modernization. While pointing abroad, *Arquitectura* argued that this was nonetheless a nationalist solution, because it dovetailed with the government's plans to "rationalize" or integrate the megalopolis to facilitate its governance and redevelopment.[30] By 1956, Pani would secure the necessary changes to the law through his extensive political connections and build his first condominium project on Paseo de la Reforma. Some years after Pani's visit to Brazil, *Arquitectura* published a profile of the Pedregulho complex in Río, excerpting Henrique Mindlin's recent book on modernist architecture there. The article emphasized the creature comforts of mass housing, including two pounds of free laundry service per person per week, and its *integración plástica*, involving collaborations with artists Candido Portinari and Roberto Burle Marx.[31] These would have particular resonance as Pani and associates prepared designs for their own massive housing complex, Nonoalco-Tlatelolco (1964) in Mexico City, which combined apartment blocks with extensive welfare and leisure facilities.

While Brazil is perhaps the best case study for considering Pani and *Arquitectura*'s articulation of a Latin American architecture, this is not to say that developments in other countries went completely unregistered. It appears that the journal's radar was tuned to other "emerging markets," where Mexican architectural expertise might be well received. Pani's clearest competitor on the hemispheric stage, at least in terms of their shared design and social interests, was Venezuelan

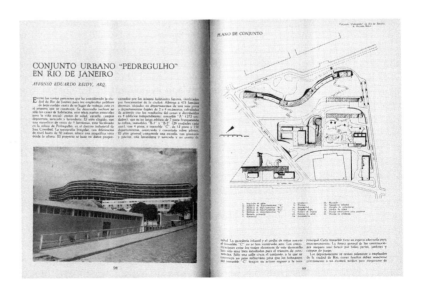

12.06 *Arquitectura*'s coverage of Affonso Reidy's Pedregulho housing complex in Rio de Janeiro, Brazil

12.07 Promotional foldout
advertisement for the
Nonoalco-Tlatelolco
housing project placed in
Arquitectura/México by the
complex's main financier

architect Carlos Raúl Villanueva. Like Pani, Villanueva came to prominence in the
1940s and 1950s through his leadership of the Ciudad Universitaria project and
mass-housing complexes like El Silencio, both in Caracas.[32] Although Villanueva
had garnered a few mentions before 1959—including five illustrated pages on his
CU[33]—the most robust coverage of Venezuela coincided with Pani's own work
there that year. A dictatorship flush with petrodollars pushed for Caracas's
redevelopment, favoring the type of high-density, mixed-used projects with which
Pani had established his reputation. He produced four plans for large-scale
commercial and residential projects, none of which were executed ultimately.[34]
According to Pani, regime change—toward democracy with the election of
Rómulo Betancourt—led to their cancellation.[35]

Like Brazil, Mexico attracted international media attention and the two were
often compared by foreign architecture critics. Pani's Nonoalco-Tlatelolco housing
complex, with over one hundred Functionalist buildings set on three superblocks,
prompted *Architectural Review* to assert: "Even after years of inoculation against
the shapes of Neimeyer, Northern Europeans can still be taken aback by Latin
American bravura."[36] Among the earliest foreign-language monographs was
Esther Born's *The New Architecture in Mexico* (1937), followed by Trent Elwood
Sanford (1947), Irving E. Myers (1952), Max Cetto (1961), and Clive Bamford Smith
(1967).[37] While this coverage cannot and should not be directly linked to
Arquitectura, Mexico's modernist architectural culture, as had its visual arts
for a long time, thrived on the injection of and evaluation by foreigners, the
impressions of its "guests." Several of these publications were enthusiastically
reviewed, and Myers and Bamford Smith would serve as correspondents to the
journal. *Arquitectura* favorably reviewed Henry-Russell Hitchcock's *Latin American
Architecture since 1945* (1955). The editors noted that Hitchcock not only recorded
important technical details of the projects featured but also, importantly, assessed
their relative "degrees of originality."[38]

For all this acclaim, Latin American modern architecture paradoxically receded
further from *Arquitectura*'s frame after the early 1960s. Pani, it appears, pulled
back from the region as did Euro-America. Why? Was he so satisfied with developments in Mexico that he no longer engaged in his malleable Latin Americanism?
Could the journal skip the hemispheric identification entirely because it was finally
plugged into international circuits with the 1968 Olympics? At the same time, a

12.08 Cover of
Arquitectura's December
1962 issue featuring the
plan of Nonoalco-Tlatelolco
as an abstract composition

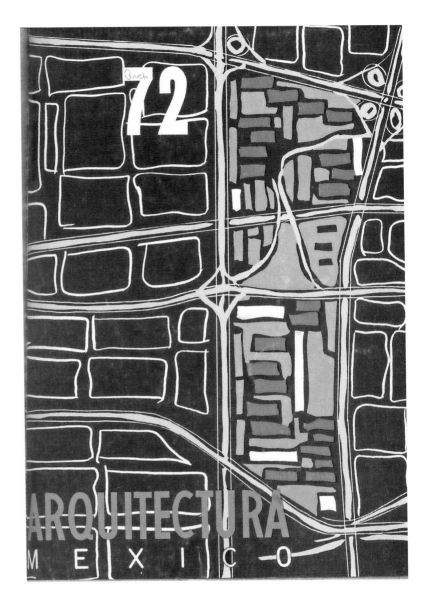

series of political and economic crises, including the Tlatelolco massacre in Mexico
and a dictatorship in Brazil, no longer made these sites utopian places that
broadcasted hemispheric solidarity. Pani's hospitality toward Latin American
modern architecture no longer served him or the journal in the same way.

Pani as Host

Pani derived much of his professional and rhetorical authority from his status as
a cosmopolitan elite. Like many of his peers, Pani spent most of the Mexican
Revolution abroad, preparing for a leading role in shaping Mexico's national
culture at its conclusion. Unlike most of his peers, whose sojourns were short
term, Pani was raised and educated in Europe. Although he was born into a
politically connected family and quickly received commissions, Pani saw himself
for a time, strategically perhaps, as situated in between Europe and Mexico.

Looking back at his career he told an interviewer, "I was neither foreign nor Mexican, I was a dubious hybrid who was either envied or thought incapable, or French-ified, not Mexican."[39] While his foreign credentials were never an impediment within an architectural and political establishment which claimed absolute sovereignty while looking studiously across the Atlantic (and the Rio Grande), such an ambivalent social identification buttressed his claim to serve as translator, and illuminates our understanding of his brand of hospitality. Through *Arquitectura* he served as host first to European modernist architecture, and then to Mexican modernism. Through *Arquitectura* he introduced recent European émigrés (including Max Cetto and Hannes Meyer) to his Mexican colleagues, often before they had arrived.[40] He also served as host to Mexico's practitioners and partisans of Functionalism, inviting contributions from Enrique Yáñez, who would head the architecture department at the Institute of Fine Arts, and José Villagrán García, Mexico's leading architectural theorist, among others. "After they adopted me, they accepted me and I was one of them," he told the same interviewer.[41]

Pani also found room in the journal for his own projects. Such self-promotion was another facet of his role as host: creating a space in which his cosmopolitan taste, personal connections, and private practice allowed him to make the necessary value judgments to contribute to an emerging architectural culture. Even as *Arquitectura* grew, serving as propaganda arm or bully pulpit to Pani's preferred projects and policy, and the architectural market moved toward industrialization and globalization, the journal cultivated the image of a domestic enterprise. Celebrating fifty issues, editor Acevedo Escobedo reminded readers that it was founded in Pani's home and was a family affair, "motivated by a generous artisanal impulse."[42] *Arquitectura* was fashioned as the natural(ized) home to Mexican architecture.

Perhaps the most productive host-guest relationship that *Arquitectura* housed was between Pani and Mathias Goeritz. Trained as an art historian, Goeritz left Germany in 1941 and arrived in Mexico (by way of Morocco and Spain) in 1949, finding colleagues among Mexico's modernist architects, who were more willing than the visual arts establishment to embrace abstraction. Goeritz's formal experiments and criticism played a critical role in propelling the *integración plástica* movement, especially its ideal of multi-disciplinary collaboration.[43] Beginning in 1959, Pani invited him to edit an arts supplement: over forty-seven issues, Goeritz and multiple contributors (especially Ida Rodríguez Prampolini) commented on currents at home and abroad, assembling the journal's largest network of ad-hoc correspondents to date.

In addition to the work of Goeritz, Pani and company followed the Spanish-language architecture journals and monographs closely, reviewing them regularly. *Arquitectura* celebrated the work of Uruguayan architect Julio Vilamajó published in a journal by the same name, this one based in Montevideo. It noted that yet another *Arquitectura*, this one in Cuba, had reprinted photographs of Pani's National Teachers College. Pani's most substantive if somewhat ambivalent contribution to the category of Latin American modern architecture was not the journal's rotating masthead, or the reproduction of material from foreign publications, but his notion of correspondence itself. Pani sought not only communication but also commensurability. In the sometimes veiled language of hospitable exchange, *Arquitectura* recognized the work of architects in Latin America and they were expected to recognize Pani and Mexican modern architecture in return.[44] Wishing *Proa* a "cordial hello and wishes of a long life" in a country with a reputation for "conservatism and academicism," the editors described the Colombian journal as an "agreeable and straightforward South

45

DEDICADA A ALGUNAS INQUIETUDES RECIENTES DEL ARTE ACTUAL

ARQUITECTURA MEXICO SECCION DE ARTE No. 28

FOTOGRAFIA MARIANA GOERITZ 1958

American magazine."[45] Closer to home, while it lauded Mexico's *Espacios*—a relatively new architecture journal that championed *integración plástica*—as "interesting and [much] discussed," it also suggested that it needed to do a better job at reining in its meandering articles.[46] *Arquitectura* also staged exchanges between Latin American colleagues in the form of invited columns. Between 1950 and 1954 it published the "Dice . . ." ("Says . . .") series that asked high-profile visitors to share their uniformly positive impressions of Mexican art and architecture. Argentine art critic Jorge Romero Brest, for example, was invited to comment on *integración plástica* a year before the Pan-American Congress.[47] Some columns were more clearly expedient: as Louise Noelle points out, providing President Miguel Alemán with a platform in 1952 coincided with the dedication of the CU, of which he was the patron (as well as Pani).[48]

Other guests veered on the edge of hospitable politesse, though never completely turning upside down the dynamic carefully cultivated in *Arquitectura*. In 1958, Ignacio Díaz Morales, an architect from Jalisco, was given space to present his interpretation of the "function of an architecture journal." While Díaz Morales's language was vague enough to suggest any journal, his criticisms would have resonated as much more specific for any reader of *Arquitectura*. He proposed more technical content for the journal's many photoessays, with material lists, greater site descriptions, and climatic and socioeconomic analysis. "Objective" photographic documentation, which had launched *Arquitectura*, was not enough in his estimation; rather, if a journal provided the data he proposed, it would avoid "the odious monotony of ephemeral international pseudo-architecture" and make it "not just another 'fashion magazine.'"[49] Although an editor's précis to an article was not unusual in *Arquitectura*, in this case it refuted the text before it was (presumably) read: Díaz Morales proposed an ideal, not a reality—and a "tedious" one at that, the editorial board argued. The précis made clear that the journal had magnanimously given Díaz Morales the space to present his views, but this would remain its space. In other words, *Arquitectura* offered hospitality in order to assert its editorial domain. Indeed, it took the opportunity to remind its readers of the valuable service it provided: "We think of our [journal] as a panorama of what is produced, and thanks to it—as it is—Mexican architects are known and appreciated abroad."

Welcoming a guest, be it a columnist or Latin America, at once depended on and helped reify *Arquitectura* as "home" to Mexican modern architecture. In the same vein, the journal played a significant role in consolidating recent Mexican architectural history. Between 1939 and 1941 it published in four installments José Villagrán García's theoretical writings, based on his lectures at the national university's architecture school. Villagrán García, who taught generations of architects, advocated a Functionalist and socially oriented practice for a resource-constricted post-revolutionary reality—a vision Pani and many of his colleagues adopted. *Arquitectura* lauded Villagrán García's Institute of Hygiene in Popotla (1925) as a point of departure for this modernist vision, and in a special issue devoted to his work, Alberto Arai, a pioneering modernist in his own right, identified him as the "pillar of contemporary architecture."[50] Such a historiography cleared room for many of the journal's own author–architects to proceed rhetorically.

In 1959, at El Presidente Hotel, Mexico's power brokers feted Pani on the occasion of his twentieth anniversary as an architect. As described in *Arquitectura*, guests included the secretary of finance and the rector of the national university. Perhaps a bit self-conscious about dedicating a whole issue to its founder and director, the letter from the editorial board posed a question it had already answered over its forty-year run:

12.09 (opposite) Mathias Goeritz's wife, Marianne, frequently provided photographs for the first page of the magazine's art section, in which construction materials take on sculptural or architectural qualities.

12.10 Mario Pani giving a
lecture at the Sociedad de
Arquitectos Mexicanos

Well, if this journal follows the norms of hospitality toward the work of others—including the case of architect Pani, when the case warrants it—then why on an occasion such as this should we not present a retrospective panorama of the architectural and urbanistic [work] realized by our animator?[51]

Or put another way, was it not the norm of hospitality that the host, on occasion, also be served? As Julián Día Arias, the sub-secretary of industry and commerce would note in his speech honoring Pani, he was much more than just an architect: without ever holding public office Pani was a "statesman."[52] The definition of the architect in Mexico was indeed changing, expanding jurisdictionally and territorially, although not always in tandem. Ultimately, Pani's influence was based not on winning commissions outside of Mexico or deliberately constructing a Latin American architecture but managing flows of local, national, and international identification and resonance within its borders, never an unfettered or disinterested enterprise.

Although Latin America is not really a tangible thing but an idea with varying degrees of currency since the nineteenth century, it is still a useful conceit in writing architectural criticism and histories. It operates as a productive field for working through the fleeting continuities and discontinuities of lived experience and social processes. What comes into focus, at least through the lens of Pani's *Arquitectura*, is that while a Latin American modern architecture may never have existed as a coherent set of monuments and discourses, hemispheric identifications, real or advantageously imagined, offered key figures in national architectural cultures a means of negotiating a globalizing architectural media and market that demanded at once universality and recognizable difference. While Pani's hospitality toward Latin America was conditional, scholarship will have to develop its own means of remaining radically open to cases—immediately visible and less so—of hemispheric travel and exchange; what Derrida suggests is an impossible though still necessary stance toward Others.

Notes

1 The journal's original title, used until 1945, was *Arquitectura: Selección de arquitectura, urbanismo y decoración*. Endnotes will refer to the journal as *Arquitectura/México*. The

Universidad Autónoma de México, under its "Raíces Digital" imprint, released a digitized version in 2008. This chapter is based on an examination of this version. All translations are the author's unless noted otherwise.

2   Because industries were protected, real estate was the more attractive investment vehicle, fueling significant speculation, especially in Mexico City.

3   There are some obvious problems in seeking to assess the status of Latin American architecture in any discourse, not least because of the slipperiness of language. In *Arquitectura/México*, "Latin America," for example, variably included and excluded lusophone Brazil. Similarly, the "universal" might signify the truly global or merely Euro-American. This chapter attempts to contextualize these terms as best as possible, but allows for their permutation.

4   As Beatriz Colomina has argued, modernist architecture cannot be isolated from—and, indeed, cannot become modern until it has engaged—mass media. Beatriz Colomina, *Privacy and Publicity: Modern Architecture as Mass Media* (Cambridge: MIT Press, 1996), 14. See also Kester Rattenbury ed., *This Is Not Architecture: Media Constructions* (New York: Routledge, 2002).

5   See Jacques Derrida, *Adieu to Emmanuel Levinas*, trans. Pascale-Anne Bault and Michael Naas (1997; Stanford: Stanford University Press, 1999); *On Cosmopolitanism and Forgiveness* (New York: Routledge, 2001); and *Of Hospitality*, trans. Rachel Bowlby (Stanford: Stanford University Press, 2000).

6   Garciela de Garay, *Mario Pani* (Mexico City: Instituto de Investigaciones Dr. José María Luis Mora, 2000), 30.

7   The photoessay was Pani's primary means of "documentation," although he did not account for the inherent bias of editorial selection, emphasis, and layout. Mario Pani, "Editorial," *Arquitectura/México* 1 (December 1938): 3–4. The editorial–administrative nucleus of the publication was initially made up of family members and close friends. Vladimir Kaspé, a friend of Pani's from architecture school in Paris, served as the journal's first "outside" editor-in-chief from 1942 to 1950. Later chief editors would include Mauricio Gómez Mayorga, Antonio Acevedo Escobedo, Antonio de Ibarrola, Arturo Pani Linaae, and Louise Noelle.

8   Pani, "Editorial," 3.

9   Georges Gromort, "La arquitectura Barroca," *Arquitectura/México* 3 (July 1939): 3–12; Gromort, "Notas sobre Palladio," *Arquitectura/México* 5 (April 1940): 3–10; Vladimir Kaspé, "Encuesta," *Arquitectura/México* 1 (December 1938): 5–20.

10 Mario Pani, "Satélite: La Ciudad fuera de la ciudad," *Arquitectura/México* 60 (December 1957): 215.

11 Mario Pani, "Una nueva etapa," *Arquitectura/México* 81 (March 1963): 3.

12 Valerie Fraser, *Building the New World: Studies in the Modern Architecture of Latin America 1930–1960* (New York: Verso, 2001).

13 M. A. Hernández, "El Colegio de Arquitectos Cubanos," *Arquitectura/México* 8 (July 1941): 61.

14 Manuel Chacón, "Arquerías," *Arquitectura/México* 13 (July 1943): 181–184. Emphasis mine.

15 See Adnan Morshed, "The Cultural Politics of Aerial Vision: Le Corbusier in Brazil (1929)," *Journal of Architectural Education* 44, no. 4 (May 2002): 201–210.

16 That is not to say that it did not sometimes realize these aspirations: foreign points of sale were listed by issue three (Bogotá, San José, Guatemala, and Paris).

17 "Facultad de medicina en Ciudad Trujillo," *Arquitectura/México* 20 (April 1946): 302–303.

18 Antonio Acevedo Escobedo, "Los 50 números de *Arquitectura*," *Arquitectura/México* 51 (September 1955): 180. Emphasis in original.

19 "Se reunirá aquí el VIII Congreso Panamericano de Arquitectos," *Arquitectura/México* 37 (March 1952): 127.

20 *Arquitectura/México* 39 (September 1952).

21 *Arquitectura/México* 28 (July 1949): 159. Cited in Carlos Flores Marini, "El debut de la Ciudad Universitaria," *Archipiélago: Revista Cultural de Nuestra América* 16, no. 60 (2008): 54.

22 Mario Pani, "La Bienal de São Paulo," *Arquitectura/México* 37 (March 1952): 78–79.

23 Pani, "La Bienal," 70.

24 Héctor Hernández, who until this point had headed the "foreign section," was integrated into the core "collaborators" staff, and served as editor.

25 Acevedo Escobedo, "Los 50 números," 180.

26 Néstor Dos Santos Lima, "Brasília: ensayo sobre la nueva capital de Brasil," *Arquitectura/México* 64 (December 1958): 221.

27 *Mario Pani, arquitecto* (Azcapotzalco: Universidad Autónoma Metropolitana, 1999), 117.

28 This transition did not occur without controversy. Obregón Santacilia took out a full-page advertisement in one of the city's major newspapers charging Pani with plagiarism. "Los planos del 'Hotel Reforma' son del arquitecto Carlos Obregón Santacilia," *Excélsior*, November 7, 1936. See also Lourdes Cruz González Franco, "Los Hoteles para un México moderno," in *Mario Pani*, ed. Louise Noelle (Mexico City: Universidad Nacional Autónoma de México, 2008), 55–67.

29 Pani founded several companies and took part in numerous speculative ventures that complemented his architectural and urban planning practices. See also Diane E. Davis, *Urban Leviathan: Mexico City in the Twentieth Century* (Philadelphia: Temple University Press, 1994).

30 "Notas y noticias: Se implanta en México el sistema de propiedad por pisos ('Condominio S.A.')," *Arquitectura/México* 44 (December 1953): 253–254; "El primer edificio en México de propiedad por pisos," *Arquitectura/México* 53 (March 1956): 3–13. See also Pedro C. Sonderéguer, "El primer edificio de condominio en México," in *Mario Pani*, ed. Noelle, 161–168.

31 H.E. Mindlin, *Modern Architecture in Brazil* (New York: Reinhold, 1956); "Conjunto Urbano 'Pedregulho' en Río de Janeiro," *Arquitectura/México* 58 (June 1957): 98–106.

32 Pani stands in contrast to Luis Barragán, and his protégé Ricardo Legorreta, who largely designed in a highly exportable "regionalist" style favored by Euro-American critics.

33 A short introduction called it "integral architecture." "Ciudad Universitaria de Caracas," *Arquitectura/México* 56 (December 1956): 217–222.

34 Enrique Molinar, "Obras de Mario Pani en Venezuela," *Arquitectura/México* 67 (September 1959): 172–176.

35 *Mario Pani, arquitecto*, 53.

36 "Mexican Triangle," *Architectural Review* 137 (April 1965): 252.

37 Esther Born, *The New Architecture in Mexico* (New York: The Architectural Record; W. Morrow, 1937); Irving E. Myers, *Mexico's Modern Architecture* (New York: Architectural Book Publishing, 1952); Max Cetto, *Modern Architecture in Mexico* (New York: Praeger, 1961); and Clive Bamford Smith, *Builders in the Sun: Five Mexican Architects* (New York: Architectural Book Publishing, 1967).

38 "Libros y revistas," *Arquitectura/México* 53 (March 1956): 62.

39 De Garay, *Mario Pani*, 45.

40 Max Cetto arrived in Mexico in 1947. His first profile in the journal was eight years before: "Escuela de cocina en Alemania," *Arquitectura/México* 3 (July 1939): 54–55.

41 De Garay, *Mario Pani*, 45.

42 Acevedo Escobedo, "Los 50 números," 179–180.

43 Jennifer Josten, "Mathias Goeritz y el arte internacional de nuevos medios en la década de 1960," in *Ready media: Arqueología de los medios e invención en México*, ed. Daniel Garza Usabiaga (Mexico City: Instituto de Bellas Artes/Laboratorio Arte Alameda, forthcoming 2012).

44 Ida Rodríguez Prampolini, "Revistas," *Arquitectura/México* 100 (April–July 1968): 110.

45 "Libros y revistas," *Arquitectura/México* 28 (July 1949): 189; "Libros y revistas," *Arquitectura/México* 29 (October 1949): 252.

46 Goeritz's arts supplement in *Arquitectura/México* began at roughly the time *Espacios* folded, carrying the torch of *integración plástica*, to which Pani subscribed.

47 Jorge Romero Brest, "Jorge Romero Brest dice. . .," *Arquitectura/México* 35 (September 1951): 258–263.

48 Louise Noelle, "La revista Arquitectura/México," in *Mario Pani*, ed. Noelle, 317–328.

49 Ignacio Díaz Morales, "Función de una revista de arquitectura," *Arquitectura/México* 61 (March 1958): 48.

50 Alberto Arai, "José Villagrán García, pilar de la arquitectura contemporánea," *Arquitectura/México* 55 (September 1956): 139–162. See also the description of the national architecture school prior to his arrival by a former student: Enrique del Moral, "Villagrán García y la evolución de nuestra arquitectura," *Arquitectura/México* 55 (September 1956): 131–132.

51 "No está por demás decir," *Arquitectura/México* 67 (September 1959): 123.

52 Ibid., 177.

13

TECHNICS AND CIVILIZATION

Félix Candela's Geopolitical Imaginary[1]

María González Pendás

On March 1, 1960, the architect Eduardo Robles wrote to his colleague and fellow-exile Félix Candela (1910–1997) to congratulate him on the publication of his work in the Spanish journal *Arquitectura*. "I was as moved as you might have been," writes Robles, "by the affection and the treatment you get from those who considered you so annoying only a few years ago, when they could not publish a thing of yours."[2] After years of conflicted relations with, and censorship of, Félix Candela, *Arquitectura*, the official publication of the Chamber of Architects of Madrid, dedicated its October 1959 issue to him.[3] This included a brief essay by Candela, an extensive thirty-two-page review of his work, and a series of laudatory commentaries by several Spanish architects and critics.[4] What was of significance for Robles, though, was less what these commentators pointed out than what they failed to mention: "they intentionally leave out something that has always characterized you: your liberal 'political' strength, in which they never believed." For Robles, this aspect was not only of paramount importance to understanding Candela's work, but was also related "to what you have done with the theories on shell construction."[5]

Robles's comment points toward two intriguing aspects of Félix Candela: that his work cannot be fully comprehended without relating it to his political discourse, and that this relationship seems most significant in the context of looking at his work—which was conceived and built in Latin America—from the viewpoint of Spain. And yet, other than a quick biographical mention of his having fought against Francisco Franco and for the Spanish Republic during the Spanish Civil War (1936–1939) and his subsequent exile from Spain during the Franquista regime (1939–1975), Candela's politics and the way these relate to his architecture are, to this day, pretty much absent from interpretations of his work. The image of Candela as a pragmatic, know-how builder and structural virtuoso has predominated since Colin Faber's influential monograph *Candela, the Shell Builder* (1963).[6] A collaborator of Candela's, Faber gave detailed technical descriptions of his concrete shells with calculations of their statics, descriptions of structural details, and plenty of construction photographs—with no hint of concerns for a larger intellectual or political discourse. A previous but lesser-known book on Candela by Félix Buschiazzo put more emphasis on Candela's role as a writer, but still failed to assess a deeper intellectual or political drive.[7] Candela certainly supported this construal of his persona as a silent builder. Shortly before his death in 1997, he admitted to not having had many intellectual preoccupations throughout his career beyond technological and building issues, and not having suffered from that "complex of the savior of humanity so common among architects, that old trope that they are going to change society and the world through architecture."[8] Architectural scholarship has largely taken this statement at face value, emphasizing a strictly formalist and technical analysis of the work, and hardly questioning Candela's politics and his status as an intellectual.[9]

However, Candela's own writings of the 1950s and 1960s radically contradict his latter disdain for the utopian bent so dear to architects. Taking on the challenge suggested by Robles in his letter, in this chapter I reveal a part of Félix Candela's political imaginary as he articulated it from exile in Mexico, and suggest a relationship between his politics and his poetics of this period. Although Candela never very directly regarded architecture as a social catalyst, not only was he concerned with devising a discourse as a public intellectual, but he also envisioned a geopolitical model that was intrinsically related to his conception of technics— and of architecture—and he eventually engaged with activist publications and institutions of the Spanish political exile. Further, it was precisely the ambiguous location that Candela himself occupied, as he looked at Spanish politics from Latin America and thought of Spain and Latin America in relation to each other, that formed the basis of his understanding of the role and the specifics of technological development in the building industry, and by extension to both his political and his architectural projects in this period. Tellingly, the one political statement that Candela explicitly endorsed through drawing was the logo for the journal of the Agrupación Europeísta de México (AEM), *Intercontinentes*, in the mid-1960s. This depicts a world globe with two sweeping arrows: one that links South America to Spain, and another linking Europe to Central and North America. As a sub-group of the intellectual elite formed of Spanish Republican exiles in Latin America, the aim of the AEM was—as imagined in the map—to launch the recuperation of Spanish democracy in joint collaboration between Spain's former colonies and the European left. Ultimately, the attempt was to offer this transcontinental assembly as a response to the bipolar ideological and economic order prevailing during the Cold War. As we shall see, it was in this in-between territory where Candela not only imagined his built structures, but also where he posited an alternative model of economic and social progress.

Discourses of Idealism: From Pan-American Utopia to Spanish Democracy

I never grow tired of proclaiming myself a Spaniard. But I don't mind being a Spaniard from Mexico. I am convinced that the future of Spain, its next renaissance, resides in these Hispanic countries of America. I don't know how I got to believe that they constitute the foundation of the next great superpower, one the world so badly needs; a world desperate under the stupid domain of the two most obtuse and materialist countries.[10]

Candela's intellectual persona was arguably born a decade after his flight from Franco's Spain to Mexico, when he simultaneously started to open up new venues for building and for writing.[11] In 1950, he built his first vaults and hypars (hyperbolic paraboloids) with his newly formed construction company Cubiertas Ala—the Experimental School in Ciudad Victoria and the acclaimed Rayos Cósmicos Pavilion in Mexico City—and published his first article—"Cubierta prismática de hormigón armado en la Ciudad de México" ("Reinforced Concrete Prismatic Shells in Mexico City") precisely in the journal that would later become *Arquitectura*, soon to be followed by his first public lecture in Mexico on the same topic.[12] These events led to very prolific years in the construction and structural development of his concrete shells, and in the presentation and publication of his lectures. These lines of work being fundamentally technical in scope, by the mid-1950s Candela started to broaden his discourse into politics more explicitly. Starting through exchanges of correspondence with colleagues in Spain and in exile, Candela began speculating about imperial power, the political and cultural future of Spain, and the relationship between Spain and Latin America. Moreover, he realized how his newly acquired status of famed architect who writes and is able to put forth his ideas in the public realm, called for rethinking his professional role by turning his discourse from technical specialism toward political commitment.[13]

AÑO II · NUMS. 8 y 9   ·   MAYO-JUNIO DE 1968

Intercontinentes

ORGANO DE LA AGRUPACION EUROPEISTA DE MEXICO

APARTADO POSTAL 6-607   LOPEZ 60 - 1º   MEXICO 6, D. F.

UNIDOS,
MEXICANOS Y ESPAÑOLES

Nos honramos publicando el discurso del Sr. Ingeniero José A. PADILLA SEGURA, Secretario de Comunicaciones y Transportes, pronunciado en nombre del señor Presidente de la República, DON GUSTAVO DIAZ ORDAZ, en el acto reseñado en otras páginas de este número.

Republicanos españoles:

Señoras y señores:

Vengo en nombre y en representación del C. Gustavo Díaz Ordaz, Presidente Constitucional de los Estados Unidos Mexicanos, a recordar la proclamación de la República Española en ocasión de su XXXVII aniversario.

Conmemoramos más que un hecho, la vivencia permanente de un ideal que nos es común: El derecho a gobernarnos libremente y a nombrar a quienes por decisión popular ejercen el poder delegado por los ciudadanos. *Sólo el pueblo tiene en todo tiempo el inalienable derecho de alterar o modificar la forma de su gobierno.*

Para nosotros, democracia y república no son meros conceptos académicos o referencias históricas: son esencia misma de la nacionalidad.

Creemos firmemente en la democracia como el sistema para la expresión de los más nobles anhelos del hombre, en su lucha permanente por superarse a sí mismo. Creemos en una república democrática, activa y profundamente humana, en la que se realiza la síntesis de ideologías y esfuerzos para lograr que todo hombre sea libre en lo político, en lo económico y en lo espiritual. Creemos en una democracia republicana íntimamente unida a las tradiciones, a las luchas, y a las esperanzas del pueblo; que se apoya en él y al mismo tiempo lo dignifica y eleva. La entendemos como sistema para lograr que los hombres vivan en paz, respetuosos de sus semejantes en la libertad y la justicia.

Por eso, republicanos españoles, aseguramos que nos unen lazos comunes, no solamente en la historia y en la sangre, sino en lo que es más importante: en el espíritu y en el pensamiento. Ustedes han sido noblemente rebeldes. Nosotros hemos sabido serlo también. En todas las circunstancias, aun en las más adversas, hemos tenido un hombre capaz de luchar tenazmente por el ideal republicano.

Nuestra nacionalidad surge apoyada en los principios de la República y en la igualdad de los hombres, y ustedes, enraizados profundamente en México, representan una idea que nos es entrañablemente cara. Pueden estar seguros que verán en este suelo, reafirmarse día a día, el concepto democrático y republicano.

La postura de México es firme, clara y terminante. No está sujeta a mutaciones y los ciudadanos que por voto popular han ejercido la primera magistratura del país, confirmaron que la doctrina mexicana es el resultado de la meditación y de los sentimientos de un pueblo que ha sufrido y luchado incansablemente por el ideal republicano.

Respetamos a quienes ajustan su conducta al mandato de sus convicciones. Los españoles republicanos lo han hecho así y por ello podemos hablar el mismo idioma en la comunidad del pensamiento democrático. Reconocemos también su contribución positiva a nuestro país, en todas las actividades humanas, fundamentalmente en las disciplinas pedagógicas, culturales y artísticas. Reiteramos que nos une la fe republicana y la decisión de consolidarla y defenderla.

Señores: el presidente Díaz Ordaz me ha encargado trasmitir a ustedes un saludo personal y el afecto del pueblo mexicano en este aniversario de la proclamación de la República Española.

El público, puesto en pié le tributó una prolongada ovación.

13.01 *Intercontinentes,* Journal of the Agrupación Europeísta de México; Candela was its first vice-president, first edited it in 1966, and also designed its logo. *Intercontinentes* 8–9 (May–June 1968)

13.02 Félix Candela's first hyperbolic paraboloid under construction. Félix Candela with Jorge Gonzalez Reyna, Rayos Cósmicos Pavilion at the Ciudad Universitaria, Mexico City, 1950

The possibility to publically leap into politics came in late 1959, coinciding with the completion of some of his most famous structures—like the Church of San José Obrero in Monterrey and the lyrical free-edge hypar for the Palmira Chapel in Lomas de Cuernavaca—when the Mexican exile group of the Federación Universitaria Estudiantil (FUE) gave a dinner in his honor. The FUE, of which Candela had been a member during his student years in Madrid, was a secular student organization that supported the Republican educational reforms against the Falangist Sindicato Español Universitario (SEU) in the early 1930s; after the Civil War it had been declared illegal and many of its members were purged. In the speech that he worked on for the occasion, Candela decided to finally present what he termed his "Thesis on Pan-Americanism."[14] Attacking many of the event's conveners, in the talk Candela refused the nostalgia that pervaded exile discourses at large, calling instead for new solutions to the political situation in Spain. New models were needed, he argued, which he was already exploring both as form-maker and as an intellectual. As he put it: "To write, to design, in a word to create has no other alternative, no other way. Why not do the same in politics? One must not look backward with nostalgia, but look forward."[15] Candela thus spoke of the "hombre nuevo" (new man) and presented a blueprint for a new geopolitical model that—just like the concrete structures that were bringing him international renown—sprang from speculation and idealism.

Primarily, for Candela, the end of Franco's dictatorship, should come neither from ideological accords within Spain, nor from claims of legitimacy from the defeated Republican government then in exile. Rather it should come from further outside, from the creation of a new political assemblage in which Spain established a renewed relationship with the Spanish-speaking countries of Central and South America. This would take the form of a new Hispanic superpower or "federation" with a new "unified government and a common set of ideals," brought together by a new meta-Hispanic identity. With this proposal, Candela sought to transcend the borders and definitions of nation-based identities with a model he defended as eminently anti-regionalist, for he considered any form of nationalism a "nostalgic sentimentalism."[16] This came as a reaction against claims of Spanishness and related nationalist cultural policies—promoted as much by prominent sections of the Republicans in exile as by Franco's government—which

13.03 Aerial view, Félix
Candela with Guillermo
Rosell and Manuel Larrosa,
Palmira Chapel, Lomas de
Cuernavaca, 1959

he considered to be the "shameful propagandistic and folkloric" image of Spain projected by both sides.[17]

The unity of this new Hispanic superpower would be based on a common language, a common history, and a new techno-economic model. In his proposal to replace the model of the nation-state with this new Pan-American federation Candela was certainly caught in an all-too familiar imperial model: the "common language" was of course Spanish and the "common history" would be a rewriting of colonial history. Besides, this new federation would initially be led by Spaniards, specifically by certain exiles in Latin America who ought to recognize their role as the "catalyst of change" and lead the "peaceful conquest" of its formation.[18] This slide into a neocolonialist discourse dangerously echoed Franco's cultural policy and ideology of Hispanism—the regime's efforts to restore closer relationships with, gain the support of, and ultimately achieve a position of hegemony over, Latin American states.[19] Candela's way around these discursive slippages was to rethink the subject of the exile, who was meant to direct the creation of this new superpower as a figure unconcerned with representing Spain and its imperial cause, as the traditional conquistador would have it. In some consonance with Jose Gaos's notion of the *transterrado*, that of the exile as having developed a crossed identity between Spain and the destination of exile, Candela's exile— himself—precisely rejected the traditional fatherland in favor of a new basis for identity.[20] As he called for at the time: "Se puede vivir sin patria" (One can live without a fatherland).[21] That is, whereas Franco based a longing for the colonial past on the very values that supported the regime at home—religion, *patria* (fatherland), and *casta* (ethnic hierarchy)—Candela argued for their near opposites —technology, rejection of *patria*, and universalism. Still, fellow-exile and historian Leopoldo Castedo cautiously tried to tame the neocolonial direction of Candela's proposal by arguing that in any possible renewed relationship with Latin America, Spain ought to act "not as a dubious mother, but as a discreet sister."[22]

Along these lines, Candela's ideas related to contemporary progressive Latin American thinkers who shifted the ideology of right-wing Hispanism, and argued against universalist tendencies in favor of the autonomy of Latin America, even if still as an extension of its colonial past.[23] The objective here, as for Candela, was to put Latin America at the forefront of the world order in the context of the Cold War. Although Candela's proposed "great third superpower" and its related meta-Hispanic social identity would eventually become universal, its most urgent task was to counter and solve the current bipolar conflict between the Soviet Union and the United States. In so doing, it would seek no alliances with, or model itself after, either of these powers. Instead, Candela devised a techno-economic model that presented an alternative both to US capitalism and to Soviet communism, at least as he understood them. A paramount aspect of Candela's vision, which directly connected it to his architectural thinking, was its foundation on a particular definition of technology, which he later termed "intermediate technology."[24] Paradoxically, and on top of the "commonality" given by language and history, the main catalyst and key to success of this prospective Hispanic superpower was its members' shared status as "weak countries," with its potential to bring together some of these countries, then at the tail end of material progress and outside of the United States–Soviet technological race. From a rear-guard position vis-à-vis technological development, this new meta-Hispanic federation could offer a model that favored feasibility, intuition, and economic and material efficiency over extreme rationalism, mass consumption, and high-tech. Efficiency and saving—"savings of time, money, and effort"—would be the motors of this new model of progress.[25] The condition of marginality from Cold War geopolitics and economics would give these countries, once united, leverage to actually occupy a central role in its resolution.

Accordingly, Candela advocated a type of humanist technology that "does not exclude man, but that provides him with simple instruments that lighten his labor without eliminating it."[26] His target was technological determinism and mechanical rationalism through which "the total mechanization of the labor of design or invention implied the elimination of man, considered a disturbing and unpredictable element."[27] Candela's take recalled contemporary critiques of rationalism, of rampant technocracy, and of the destructive power of technology, as crudely epitomized by the Second World War and all-too ubiquitous during the Cold War. Arguments along these lines in architecture by Sigfried Giedion and Lewis Mumford were of specific relevance for Candela.[28] Mumford had already called for discerning the imprint of the human hand in mechanical work, for the machine to "restore the delicacy and sensitivity of the organism."[29] Much less sophisticated in his argument and without a historical narrative, Candela, like Mumford, advocated for a "third stage" of technological development, which Mumford had termed the "neotechnic phase," where a new social order would develop by reorienting technology toward an organic relationship with man.[30]

Candela's audience in 1959 responded badly to his critiques of the *patria*. The talk was nevertheless an effective political launch that enabled him to establish alliances with the younger and more activist exiles, especially Francisco Giral, son of the former president of the Republican Government in exile, and an influential figure in Mexican academia and intellectual circles. Although Candela did not follow Giral's project of resistance to Spanish politics all the way to the 1970s, the pairing of Candela with Giral in 1959 is significant for two reasons: to understand the radical tone of his message at this point in time, and to emphasize Candela's belief in the role of the architect as a man of science, and of the scientist as a technical professional *and* an intellectual on call for political action.[31] More specifically, Giral brought Candela close to members of the international left and the Republican government in exile who would work in the following years toward the founding of the Agrupación Europeísta de México (AEM). The Agrupaciones Europeístas were part of the Movimiento Europeo based in Brussels and of the Consejo Federal Español based in Paris, and aimed at uniting the intellectual elite of Spanish exiles and bringing them to collaborate with the European left, in an associated effort to recuperate Spanish democracy by advocating for the formation of a European federation, articulated as *Europeísmo*.

In the summer of 1965, Candela traveled to Paris, London, New York, and Washington and met prominent leftists like the leader of the Spanish Partido Obrero de Unificación Marxista (POUM) and editor of *Mañana* Julian Gorkin, as well as Russian émigré architect Berthold Lubetkin, with the objective, as he put it, of campaigning for a "solution to the Spanish problem."[32] The trip certainly allowed him to gain the confidence of the political elite in Mexican exile. Upon his return to Mexico, Manuel Torres Campaña, former Minister of the Spanish Republic and President of AEM, invited Candela to join the organization and then named him first vice-president.[33] The objective of the Mexican section was to include Central and South America in this Europeanist democratic project—thus expanding its geography across the Atlantic—and to place culture at its center. The basic objectives of the group were to create a Hispano-American cultural commission and to promote educational reform in Spain. Immediately after his affiliation, and following this line of favoring education and culture, Candela promoted the journal *Intercontinentes*; the only one published by the Agrupación Europeísta; the first issue announced its purpose as "promoting Democratic Constitutionalism in the American Republics," as well as in Spain, from the platform of those countries which had "already solved that problem," meaning Chile and Mexico.[34] This intercontinental alliance was well attuned to Candela's early vision of a Pan-American federation—the one sketched for the logo of the journal.

During the second half of the 1960s, Candela gained the confidence of the opposition to Franco, both in exile and in Spain, and his political writings and associated work as an editor, as well as his political initiatives, increased substantially. His most significant piece was the quasi-manifesto "España, democracia, constitución" ("Spain, Democracy, Constitution") published in Victoria Kent's magazine *Ibérica: por la libertad* and in an appendix to the second issue of *Intercontinentes*.[35] In it he articulated the AEM's call for constitutional democracy. Despite the journal's evocative title, Candela's earlier ideas on an expanded geography became blurred as he called for national integration and consensus within Spain, appealing to the

> right to collaborate patriotically for the advent of Constitutional Democracy: to those from above, those from below, and those from the middle; the ones from the right, the left, and the center. Spain is not Spain anymore if she is not constituted by all of her sons.[36]

In order to fit with AEM, and as they moved away from their initial intercontinental projections, Candela gradually departed from his advocacy of a new Hispanic coalition and took a more pragmatic tone with regard to the call for democracy in Spain.

Ibérica
POR LA LIBERTAD

En este número

ESPAÑA
DEMOCRACIA
CONSTITUCIÓN
Félix Candela

LA ÚLTIMA CRISIS
Gregorio López Cid

CARTA AL DIRECTOR
GENERAL DE INFORMACIÓN
Juan García Durán

EDITORIALES
Escupir al cielo
La prensa en "libertad condicional"

SIN PERMISO DE LA CENSURA
Parálisis oficial
Atentado contra la inteligencia

RESUMEN DE NOTICIAS

VOLUMEN 13, NO. *precio 25c*   **15 DE SEPTIEMBRE, 1965**

13.04 Exile publication edited by Victoria Kent with an article by Candela and excerpts of his letter declining the invitation to return to Spain in 1964. *Ibérica: por la libertad* 13, no. 9 (1965)

13.05 Cover of "Temas de hoy," appendixed flyer to *Intercontinentes* 2 (August 1966), with Candela's article "España, democracia, constitución," alongside other pro-Europeanist essays

TEMAS DE HOY

España
Democracia
Constitución
por Félix Candela

Europeísmo
por E. López Sevilla

España en Europa
por "Juan Bizcaino"

FOLLETO No. UNO

Suplemento al Nº 2 (Agosto)

de

"INTERCONTINENTES"

AGRUPACION EUROPEISTA DE MEXICO

1966

This reorienting of discourse toward the advocacy of patriotism and of consensus was paradoxically attuned to his earlier rejection of *patria*. The essay appeared accompanied by a political statement of a different sort, one more closely related to architecture. In late 1964, the director of the journal *Arquitectura*, Carlos de Miguel, invited Candela to give a lecture for the opening of an exhibition on Antonio Gaudí organized for EXCO (Exposición Permanente de la Construcción) in Madrid and under the auspices of the Ministry of Housing. Candela rejected the invitation in a letter to de Miguel where he expressed the "personal conflict" the invitation had prompted: "I am deeply sorry, but I cannot decently be a guest of the Spanish Government."[37] Candela made this refusal the spearhead of his public politics during these years, publishing parts of the letter and sending copies to various colleagues in AEM.[38] At this point in time, Candela resisted physical return to the *patria*, a resistance he admitted was idealistic, but relevant precisely by virtue of this very idealism. Only at a distance from Franco's Spain could one still maintain a political conscience:

> We live in safe times, but there were moments in history when people were killed for the improbable causes that invoke the basis for life in society. Even at the risk of falling into ridicule, a certain dose of idealism can only be fomented and sustained by Quixotic acts, which are those acts where the author, in his craziness, believes he is heroic. Those of us living outside Spain are in a privileged position since we are not forced to give in to a series of small, personal betrayals. Individually considered, these betrayals may seem of little relevance. But taken as a whole, they tend to degrade collective life.[39]

The reaction to this letter from the representative of the Republic in the United Nations and key figure in AEM, Manuel Díaz-Marta, echoed one of the main conflicts for Franco exiles, the fact that invitations to return and the related image of openness and tolerance was the new means for Franco to gain legitimacy:

> Invitations and flattering events for Spaniards who had triumphed [abroad] are part of the task the dictator and his followers set themselves some time ago as a means to eliminate the enemy. Initially it was physical suppression, the lack of freedom, prosecutions, and even exile. Now the tactics have changed.[40]

Full return meant simply giving up the whole project of exile. It would mean the end of idealism, the beginning of the end of politics, or as Candela put it, the "beginning of the degradation of collective consciousness."[41]

Shells of Idealism, or on Candela's Politics and Poetics

How did Candela's pervasive political thinking and his status as an intellectual imbue, and in turn become influenced by, his architectural work? In his letter of March 1960, as elsewhere, Robles was never very explicit as to the connections between Candela's work and politics. Candela himself was quite cryptic on this difficult point. In the 1959 FUE talk, he specifically related political thinking and structural design, but only in terms of a common attitude of innovation and idealism. Just as he had reacted against traditional theories of structural design for the development of his signature shell construction system, he set himself "to do the same" in politics, and to develop a political project that sprang from a similar "critical spirit" and was a fight against the status quo of rationalist and nationalist ideologies.[42]

Besides a question of attitude, the connection between the poetics and the politics in Candela coalesced around two main points. One was his conception of technics, or what he termed technology, but went beyond mere utilitarian or material aspects and into a broader conception of social and cultural methods and dynamics. For Candela, no formal or stylistic a priori directed his architecture. Instead, technology led directly to the definition of the built object, with aesthetics and function alongside it. Following Mumford, for Candela technics also originated a social contract: "[Technics] implies a process of civilization, or just as much, a process of renunciation of freedom and individual rights when these are in conflict with social objectives."[43] Technics, in the form of concrete shell construction and geometric structural thinking, was for Candela not only a basis for his building system, but also a defining force in negotiating "political space," and eventually in providing a new model of progress. Moreover, the building industry—at least as he practiced it in Mexico in the 1950s—provided the ideal realm for that "intermediate technology" in which the new Hispanic superpower would base its progress. Architecture and construction in this context offered an "underdeveloped" position of resistance to the "technology innovation and industrialization" rampant elsewhere.[44] They thus offered the disciplinary model for an economic development that favored material efficiency over consumption, intuition over reason, and humanism over materialism. His shell structures deployed all these qualities, he argued: efficiency in the sense of using the least amount of material possible or the thinnest concrete section, for the definition of the most space; intuition in the sense that the construction required more hands-on trials than mathematical calculation; and humanism in terms of a building industry close to craftsmanship, with a limitation on the mechanization that embodied the destructive and alienating power of mass technification. Just as his structures were governed by the "natural laws of balance and economy," he was hoping for sociopolitical balance through a technological and material economy of efficiency.

The contradictions inherent in Candela's vision become all the more evident here. The efficiency that he called for operated in the initial stages of design and calculations, and in the finished product. But the construction process entailed an extreme deployment of material—wood for the framework—and of labor. Economic efficiency was only a plausible argument while the cost of wood remained low, and because his shell construction method required large amounts of unskilled labor at a moment when minimum wages and unions were still not operative in Mexico. Thus Candela's idealization of humanist technology was close to man only to the extent that the particular man responsible for it—the worker—was not quite able to articulate his rights. On this point, Candela's discourse loops back to his rejection of *patria*, that is, to his statement that one could live without a fatherland and that a new sociopolitical order could be conceived without national boundaries. But it is a loop back that renders this element of his thought as inconsistent, and closer—if unwittingly—to the most reactionary claims of *casta* and politics of Hispanism. Candela's technical model was in the end overlaid with the politics of race and class, as non-unionized construction workers were largely of indigenous, pre-colonial ethnic origins. For all the intent to do away with nation-based identity, the class/race hierarchies operative in Mexico at the time—as in the majority of the technically "backward" countries he summoned—remained intact in Candela's vision.

A second point should be made on a more metaphorical level, and relates to his idea of territory. The structural rationale for Candela's concrete shells—mainly hyperbolic paraboloids, or hypars—was based on the static equilibrium acquired by geometrical form over the specifics of the material. That is, their stability came from the geometry: saddle-shaped, ruled surfaces with double curvature, ideally boundless, and generated by the succession of straight lines. These were static when thought of as abstract and extensive geometries, from which Candela would isolate a fragment. This was easily constructed by using wooden planks for the formwork of poured-in concrete. The result was "limitless forms, forms in continuous expansion," as Fernado Chueca put it, in striking contrast to the Euclidean, limited, and finished forms more proper to orthogonal structures.[45]

Somewhat analogously, his political model was based on an "amplified patriotism."[46] Candela imported—arguably with little reflection—a spatial logic based on extensive thinking and universalism into his political model. As a result, Candela's spatial and formal referents are somewhat generic. Although these constructions were indeed only possible as products of the Mexican tradition of building with concrete, and of the construction costs and labor conditions of the time, at a symbolic level Candela's shells specifically rejected localism. They were *of* Mexico, but not necessarily willing to represent it aesthetically or historically. This becomes clearer when compared with attempts by some of his contemporaries to define a locally grounded modernity. For instance, Juan O'Gorman was at the time calling for modern architecture to become "realist, national, and regional" through explicit links with "formal references and material usage," as well as with technique connected to pre-colonial times, in the same way claimed by Candela. In this way, "the people" would identify it "as their own . . . and become a human

13.06 Wood formwork for hypar under construction. Félix Candela with Guillermo Rosell and Manuel Larrosa, Palmira Chapel, Lomas de Cuernavaca, 1959

13.07 Workers. Félix
Candela with Enrique de la
Mora, Stock Exchange,
Mexico City, 1955

manifestation of a distinct culture."⁴⁷ These linguistic and aesthetic aspirations
for a local character and style were far from Candela's concerns. If one likened
the stone mosaic murals of O'Gorman's Central Library for the Ciudad Universitaria
to Candela's analysis of vaults, the disconnection of the latter with any particular
culture, context, and site further comes to the fore. This was the point he was
making in his FUE 1959 talk with his rejection of *patria* and folklorism, and with
his new Hispanic federation, as flawed as this was. Ultimately, Candela attempted
to avoid any symbolic representation of national or regional ideologies, in order
to define an Esperanto-like architecture, a non-local, unifying, and universally
valid space; just as he argued for his new Pan-American identity.

In addition, one of the most problematic aspects of Candela's shells was their
weakness in connection to the ground, the discontinuity between the logic of the

13.08 (top) Félix Candela shell under construction

13.09 (below) Hand-drawn sketch of hyperbolic paraboloid showcasing the load-bearing system based on surface geometry, and with a note by Candela concerning isolating a section for construction. Félix Candela, November 1952

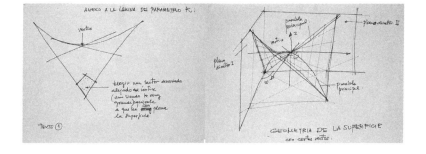

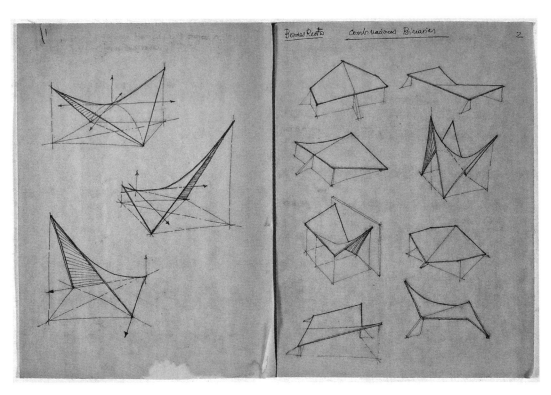

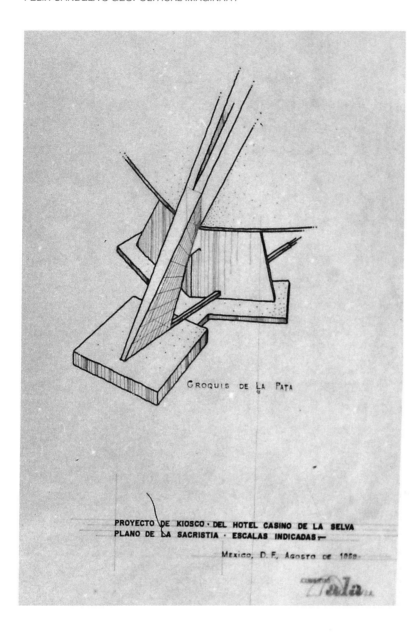

CROQUIS DE LA PATA

PROYECTO DE KIOSCO · DEL HOTEL CASINO DE LA SELVA
PLANO DE LA SACRISTIA · ESCALAS INDICADAS :—

MEXICO, D. F., AGOSTO DE 1952.

13.10 (facing top) Wood formwork for hypar under construction, showing the succession of straight lines that form the hypars. Félix Candela with Guillermo Rosell and Manuel Larrosa, Palmira Chapel, Lomas de Cuernavaca, 1959

13.11 (facing below) Hand-drawn perspectives of hyperbolic paraboloids: on the left, Candela's disregard for context in relation to his structures; on the right, the awkward and problematic nature of their supports or legs. Félix Candela, undated

13.12 (left) Detail showing the complexity of solving the support for Candela's otherwise simple structures. Félix Candela, Casino de la Selva Hotel Project, Cuernavaca, 1952

shell and the way it was supported. Candela himself admitted that the supports were key, but supplementary to his structures: "the success of most of my work stems primarily in the detailing of the legs or supports, and in how discreet and concealed they are . . . something that has not much to do with the shell itself."[48] Added to the fact that the buildings were often located outside of the urban center, or at least did not require a relationship to other buildings, Candela's shells often seem disconnected from the site, like alien objects that have just landed on location Could one see these placeless shells as the formal counterparts to Candela's political rejection of any link to a homeland?

Candela's eventual and much-resisted physical return to Franco's Spain—he finally visited the country in late 1969, some six years before Franco's death—and the

13.13 Exterior view, Félix
Candela and Fernando
Alvarez Ordoñez, Los
Manantiales Restaurant,
Xoichimilco, 1958

symbolic acceptance of the fatherland this implied, marked in fact the end of his
political commitment. This was also the twilight of his experimental constructions:
in 1968, Candela had built his last structure, the Olympic Pavilion in Mexico City;
in 1969, he withdrew from his construction company Cubiertas Ala; in 1971,
Candela moved to the United States, shifted his career toward pedagogy and
never again built large-span structures.[49] By then, Candela seems to have lost
both his political and his poetic idealism. In fact, Candela lost his acute awareness
of the bridge between technology and politics, since he agreed to return to speak
at a congress for the International Shell Association due to the association's
apparent lack of connection to the Spanish government.[50] In the end, and as
Diaz-Marta predicted, Candela was received and celebrated by government
officials. Along with the shifting economics of the 1970s, which meant a rise in
the cost of wood and in workers' wages, which played a significant role in
diminishing Candela's practice, the fact is that the endgame of his political
discourse and of his experimental period in building came in tandem, and
coincided with his eventual embrace of *patria* in Spain.

The *Intercontinentes* logo map that Candela supported at the height of his political
and structural speculations was not only at the basis of his thinking. In a sense, he
also embodied it. Whether he is to be considered a Latin American architect, a
Spanish engineer, a Mexican building contractor, or none, or all, of the above is still
a point of scholarly discussion. It has already been suggested that Candela's career
must be analyzed in relation to the sense of *desarraigo* (uprootedness), provoked
by the experience of exile.[51] Rather than fueling the debate on Candela's identity,
or his longing for it, his lack of a secure location appears as the point of departure
for reconstructing his full-fledged political imaginary. Candela's ambiguity regarding
his identity was a constituent aspect of his architecture *and* of his political thinking,
which this chapter reveals only in part. A close look at Candela's private and public
writings of the time discloses an outspoken figure who extended the base of his
architectural work into a political project. His rejection of fatherland was in essence
a claim of statelessness that came from his location between Latin America and
Spain; as consequence of his long-lasting refugee status in Mexico, and of his
problematic relationship with and conflicted return to Spain and its architectural

culture. As ambiguous as this point was in his thought, Candela folded it into his political project as much as into his architectural thinking. More than other prominent members of the Spanish architectural exile in Latin America, Candela consciously and intensively pursued politics, reacted to the current bipolar world order of the Cold War, and extended his thinking into larger social concerns. The relationship between form and politics in Candela, as elsewhere, is certainly complex and anything but binding. And yet, one could not have existed without the other. In many ways, Candela aimed for no less than Hannes Meyer's call in 1926 to "undermine the local concept of the 'homeland.' The fatherland goes into a decline. We learn Esperanto. We become cosmopolitan."[52] Both his built work and his writings in the 1950s and 1960s—in a word his discourse—read as a critique of the nation-state and of rationalism, as an unapologetic and quintessentially modern rejection of roots, if with humanist undertones, and as a need to envision a new model of identity and an alternative model of socio-economic progress. These ideas were intrinsically related to his particular take on technology and to his attitude toward structural design and form-building. This led to the conception of a new architecture—new in the sense that it was innovative in technological and aesthetic terms—that resisted anything symbolic and that strove for a certain spatial universality, and that his sociopolitical vision at one point likewise imagined.

13.14 Detail of support, Félix Candela and Fernando Alvarez Ordoñez, Los Manantiales Restaurant, Xoichimilco, 1958

Notes

1   Thanks to Jorge Francisco Liernur for his comments on a previous version of this chapter, and to Angela Giral for her insights. Research for this chapter was made possible by the support of the Fulbright Commission, the Fundación Caja Madrid, and the Graduate School of Architecture, Planning and Preservation at Columbia University, New York.

2   Eduardo Robles to Félix Candela, March 1, 1960. Box 05.09, Félix Candela Architectural Records and Papers Collection, Avery Drawings and Archives, Columbia University, New York (hereafter FCA-CU). Original in Spanish. All translations from Spanish by the author unless otherwise noted.

3   I have fleshed out Candela's political vision as it developed in the 1950s and 1960s and in reaction to his conflicted relationship to Spain and to its architecture culture in "Geometría Extensiva también como proyecto político," in *Miradas Cruzadas, intercambios entre Latinoamérica y España en la Arquitectura Española del siglo XX* (Pamplona: T6 Ediciones, 2008), and "Apátridas Architectures: Candela, Sert, and the Return of the Modern to Postwar Spain" in *Coming Home?* vol. 1, *Conflict and the Return of Spanish Civil War Exiles*, ed. Scott Soo and Alicia Pozo-Gutierrez (Newcastle: Cambridge Scholars Press, forthcoming).

4   Carlos de Miguel, ed., "Láminas de hormigón armado: Félix Candela," *Arquitectura* 10 (1959): 2–33.

5   Robles to Candela, March 1, 1960.

6   Colin Faber, C*andela, the Shell Builder* (New York: Reinhold, 1963). Similarly see Luigi Margani, *Le Superficie Rigate nelle Strutture Architettoniche* (Catania: Università di Catania, 1965).

7   Félix Buschiazzo, *Félix Candela* (Buenos Aires: Instituto de Arte Americano e Investigaciones Estéticas, 1961).

8   Félix Candela, "Discurso de aceptación al premio del Colegio de Arquitectos e Ingenieros de Madrid," *Arquitectura* 303 (1995): 46.

9   See David Billington and Maria E. Moreyra, *Félix Candela: Engineer, Builder, Structural Artist* (New Haven: Yale University Press, 2008); Enrique X. de Anda, *Félix Candela, 1910–1997: The Mastering of Boundaries* (New York: Taschen, 2008); Miguel Seguí, ed., *Félix Candela, arquitecto* (Madrid: MOPU 1994); Miguel Seguí, *Félix Candela y Emilio Pérez Piñero: Un diálogo imaginal* (Madrid: Rueda, 2004); Juan Antonio Tonda, *Félix Candela* (Mexico City: Consejo Nacional para la Cultura y las Artes, 2000). Salvador Perez Arroyo briefly mentions Candela's critique of Franco's regime during the 1960s, but otherwise insists on the image of Candela the builder: Salvador Perez Arroyo, *Los años críticos: 10 arquitectos españoles* (Madrid: Fundación Antonio Camuñas, 2003), 28. Two recent monographs start to open up Candela's personal and ideological history, but largely leave aside the specifics of his politics: Pepa Casinello, ed., *Félix Candela. Centenario 201: la conquista de la esbeltez* (Madrid: Fundacion Juanelo Torriano, 2010), and Juan Ignacio del Cueto, ed., *Félix Candela 1920–2010* (Madrid: Sociedad Estatal de Conmemoraciones Culturales, 2010). In the latter, Candela's activist role in Spain during the last years of the Republic is well addressed by Inés Sanchez de Madariaga.

10 Félix Candela to Matilde Ucelay, August 7, 1957. Box 03.14, FCA-CU.

11 Cueto, ed., *Félix Candela 1920–2010*, 75–77. Lázaro Cárdenas's politics and policies made Mexico the preferred destination of exile for Spanish Republicans in Latin America. For an account of the architects who went into exile, see Henry Vicente Garrido, et al., *Arquitecturas desplazadas: Arquitecturas del exilio español* (Madrid: Ministerio de Vivienda, 2007); specifically for the case of Mexico, see Juan Ignacio del Cueto, "Arquitectos españoles exiliados en México: Su labor en la España republicana (1931–1936) y su integración en México" (PhD diss., Universidad Politécnica de Catalunya, Barcelona, 1996). Exile scholarship from the point of view of architecture remains biographical and contains little interpretation of the exiles' politics, which, as I argue elsewhere, comes more clearly to the fore when focusing on the figure of return over the experience of exile per se.

12 Félix Candela, "Cubierta prismática de hormigón armado en la Ciudad de México," *Revista Nacional de Arquitectura* 99 (1950): 126–32. For an account by Candela of the structures mentioned here, see *Arquitectura* 303 (1995): 46–56.

13   See correspondence with Enrique Segarra, Eduardo Robles, and Matilde Ucelay, former professional and political colleagues of Candela in Spain. Box 03 and others, FCA-CU.

14   The talk before FUE can only be reconstructed from schematic manuscript notes by Candela, hereafter FUE talk notes, FCA-CU. I was not able to find a formal, structured text prepared for the occasion. I date this event to late 1959 based on a series of letters between Eduardo Robles and Candela.

15   Candela, FUE talk notes.

16   Ibid.

17   Candela to Ucelay, August 7, 1957. FCA-CU.

18   Candela, FUE talk notes.

19   Lorenzo Delgado Gomez-Escalonilla, *Diplomacia franquista y política cultural hacia Iberoamérica, 1939–1953* (Madrid: CSIC, 1988), 56.

20   Antonio Monclús, "José Gaos y el significado de Transterrado," in *El pensamiento español contemporáneo y la idea de América* (Barcelona: Anthropos, 1989), 34.

21   Candela, FUE talk notes.

22   Leopoldo Castedo to Félix Candela, August 8, 1965. FCA-CU.

23   Ricardo Perez Monfort, *Hispanismo y Falange: los sueños de la derecha Española y México* (Mexico City: Fondo de Cultura Economica, 1992), 24.

24   Félix Candela, "Influencia de la tecnología en la creatividad arquitectónica," *Arquitectura* 303 (1995): 66. Typewritten drafts of this essay exist in several languages. Box 29.13, FCA-CU.

25   Félix Candela, "Arquitectura: Arte a extinguir," typewritten original, 1. Box 05.09, FCA-CU.

26   Candela, "Influencia de la tecnología," 66.

27   Ibid., 63.

28   Candela, "Influencia de la tecnología," 60. An interesting referent here for Candela was Jaques Ellul, whose Catholic critique of technology he would have encountered later.

29   Lewis Mumford, *Technics and Civilization* (1934; New York: Harbinger, 1963), 359.

30   Ibid., 434.

31   Francisco Giral, *Ciencia española en el exilio, 1939–1989: El exilio de los científicos españoles* (Madrid: Centro de Investigación y Estudios Republicanos, 1994).

32   Félix Candela to Leopoldo Castedo, July 21, 1965. Box 02.15, FCA-CU.

33   Manuel Torres Campaña to Félix Candela, July 24, 1965. Box 14.24, FCA-CU.

34   *Intercontinentes* 1, 1. Also in "Estatuto de Intenciones de AEM." Box 14.24, FCA-CU.

35   Félix Candela, "España, democracia, constitución," *Ibérica por la Libertad* 13, 9 (September 1965): 3–4; and "Temas de hoy," appendix to *Intercontinentes* 2 (August 1966). The reference here is to the original typewritten manuscript in Box 02.15, FCA-CU. Although signed by Candela, the text could have been written in collaboration with other members of the Agrupación. Other political writings of the period by Candela are "El constitucionalismo," *Ibérica* (November 1966): 7; and "Dictadura no, democracia sí," *Ibérica* (April 1967): 4–5. The arguments here follow from those made in "España, democracia, constitución."

36   Candela, "España, democracia, constitución."

37   Félix Candela to Carlos de Miguel, December 12, 1964. Box 03.17, FCA-CU.

38   *Ibérica* 13, 9 (September 1965): 4; *Tiempo* (October 25, 1965): 62–63.

39   Ibid., 63.

40   Manuel Díaz-Marta to Félix Candela, February 1, 1965. Box 02.15, FCA-CU.

41   Félix Candela, manuscript notes for "Conferencia sobre el exilio." Box 03.17, FCA-CU.

42   Candela, FUE talk notes.

43   Candela, "Influencia de la tecnología," 60.

44   Ibid., 65.

45   Fernado Chueca, "Comentario," in *Arquitectura* 10 (1959): 22.

46   Candela, FUE talk notes.

47   Juan O'Gorman, "En torno a la integración plástica," *Espacios* (July 1953); rpt. in Ida Rodríguez Prampolini, *Juan O'Gorman: Arquiteco y pintor* (Mexico City: Universidad Nacional Autónoma de México, 1982), 93.

48   Félix Candela, "Letter," *Arquitectura* 10 (1959): 2. Alberto Pérez-Gómez has pointed toward a similar argument in Edward R. Burian, *Modernidad y arquitectura en Mexico* (Barcelona: Gustavo Gili, 1998), 46.

49 Cueto, "El ocaso de los cascarones," 104–106.

50 In letter from Manuel Solà-Morales to Félix Candela of January 30, 1968, Solà-Morales invited Candela to attend the colloquium "Industrialización de la Construcción" at the Colegio de Arquitectos in Barcelona, an invitation that Candela likewise rejected. For the details of Candela's trip to Spain, see the exchange of letters between Candela and Florencio del Pozo in 1968 and 1969, in the FC archives.

51 Miguel Angel Baldellou, "Desarraigo y encuentro," *Arquitectura* 303 (1995): 16–19; Ricardo Aroca, "A modo de presentación," in Miguel Segui, ed., *Félix Candela, arquitecto* (Madrid: MOPU, 1994), 13.

52 Hannes Meyer, "The New World," in Claude Schnaidt, *Hannes Meyer: Bauten, Projekte und Schriften—Buildings, Projects, and Writings* (Teufen: A. Niggli, 1965), 91.

IMAGE CREDITS

INT.01 Francisco Bullrich, *New Directions in Latin American Architecture* (New York: George Braziller, 1969). Reprinted with the permission of George Braziller, Inc.

INT.02 Francisco Bullrich, *Arquitectura latinoamericana, 1930–1970* (Buenos Aires: Editorial Sudamericana, 1969)

INT.03 Sylvester Baxter, *Spanish-Colonial Architecture in Mexico* (Boston: J. B. Millet, 1901). Avery Architectural and Fine Arts Library, Columbia University

INT.04–INT.05 Manuel F. Alvarez, *Las obras de arquitectura en la América Latina y en los Estados Unidos de América* (Mexico City: Secretaria de Comunicaciones y Obras Públicas, 1921)

INT.06 Martín Noel, *Contribución a la historia de la arquitectura Hispano-Americana* (Buenos Aires: Talleres S. A. Casa J. Peuser, 1921). Avery Architectural and Fine Arts Library, Columbia University

INT.07 *Arquitectura/México* 6 (July 1940)

INT.08 Walter Gropius, ed., *Internationale Architektur* (Munich: A. Langen, 1925). Avery Architectural and Fine Arts Library, Columbia University

INT.09 Alberto Sartoris, *Gli elementi dell'architettura funzionale, sintesi panoramica dell'architettura moderna*, 1st ed. (Milan: Ulrico Hoepli, 1932). Avery Architectural and Fine Arts Library, Columbia University

INT.10 Alberto Sartoris, *Gli elementi dell'architettura funzionale, sintesi panoramica dell'architettura moderna*, 2nd ed. (Milan: Ulrico Hoepli, 1935). Avery Architectural and Fine Arts Library, Columbia University

INT.11 *Architectural Record* (April 1937), Architectural Record © 1937, The McGraw-Hill Companies

INT.12 © 1943 The Museum of Modern Art, New York

INT.13 © 1955 The Museum of Modern Art, New York

1.01, 1.03, 1.05, 1.09–1.10, 1.12 Courtesy Archive Casa de Lucio Costa

1.02 J. D. Vidal, ed., *Lucio Costa: desenhos de juventude* (Belo Horizonte: Companhia Brasileira de Metalurgia e Mineração, 2001). Courtesy Biblioteca Paulo Santos

1.04, 1.06, 1.11, 1.13–1.14 Arquivo Central do IPHAN—Seção Rio de Janeiro

1.07 Arquivo Gustavo Capanema, Centro de Pesquisa e Documentação de História Contemporânea do Brasil da Fundação Getúlio Vargas, Rio de Janeiro

1.08 Biblioteca de Arte, Fundação Calouste Gulbenkian, Lisbon

2.01 Centro de Documentación e Información del Instituto de Teoría de la Arquitectura y el Urbanismo, Facultad de Arquitectura, Universidad de la República, Uruguay; Photograph by Jorge Nudelman Blejwas

2.02, 2.06–2.07, 2.09–2.10 Francisco Nogueira Martínez, *Doctor Honoris Causa Carlos Gómez Gavazzo, Profesor Arquitecto Urbanista* (Montevideo: Instituto de Teoría de la Arquitectura y el Urbanismo, Facultad de Arquitectura, Universidad de la República, 2002)

2.03–2.04 *CEDA* 6 (July 1934)

2.05 *Arquitectura* 188, no. 1 (1937)

2.08, 2.12–2.15 Centro de Documentación e Información del Instituto de Teoría de la Arquitectura y el Urbanismo, Facultad de Arquitectura, Universidad de la República, Uruguay

2.11 Courtesy Justino Serralta

2.16–2.18 Justino Serralta, *L'Unitor* (Paris: Justino Serralta, 1981)

3.01, 3.06 Bruno Zevi Foundation

3.02, 3.04 Photographs by Fabrio Grementieri

3.03, 3.05 Photographs by Leonardo Codina

4.01 *El Arquitecto Peruano* 13, no. 146 (September 1949)

4.02 PARS International Corporation, New York

4.03 *El Arquitecto Peruano* no. 279–281 (October–December 1960)

4.04–4.05 Fernando Belaúnde Terry, *La Conquista del Perú por los Peruanos* (Lima: Editorial Tawantinsuyu, 1959), 22, 112

4.06 Fernando Belaúnde Terry, *Pueblo por Pueblo* (Lima: Ediciones "Tawantinsuyu," 1960), 101

4.07 Ephraim George Squier, *Peru: Incidents of Travel and Exploration in the Land of the Incas* (New York: Hurst & Company, 1877), 555

4.08 *El Arquitecto Peruano* no. 285–287 (April–June 1961)

4.09, 4.16 Archivo Pedro Ramírez Vázquez, Mexico City

4.10 Manuel Amábilis, *El Pabellón de México en la Exposición Ibero-Americana de Sevilla* (Mexico City: 1929)

4.11–4.15, 4.17 Photographs by Luis Castañeda

5.01–5.02, 5.04–5.05, 5.07–5.11 Villanueva Foundation Archives

5.03, 5.18 Urban-Think Tank

5.06 Personal archive of Carlos Celis Cepero

5.12 *L'Architecture d'aujourd'hui* 88 (1960)

5.13 *Town Planning Review* 31 no. 3 (October 1960)

5.14 Charles Abrams, *Man's Struggle for Shelter in an Urbanizing World* (Cambridge: MIT Press, 1964), 53. © 1964 Massachusetts Institute of Technology, by permission of the MIT Press

5.15 John F. C. Turner, *Housing by People: Towards Autonomy in Building Environments* (London: Marion Boyars, 1976)

5.16 © Sabine Bitter and Helmut Weber, from the series "Caracas, hecho en Venezuela," 2003

5.17 Urban-Think Tank/Kulturstiftung des Bundes, Caracas Case Project

6.01–6.04, 6.06–6.09, 6.12, 6.14 Archives of Miguel Lawner, José Medina, and Sergio González

6.05 *Hechos Mundiales* magazine, Editorial Nacional Quimantú, 1972

6.10 *AUCA* 22 (1972). Archives of Miguel Lawner, José Medina, and Sergio González

6.11 Archives of the Ministry of Foreign Affairs, Chile

6.13 *El Mercurio*, September 13, 1972

6.15 Photograph by Marcelo Montecino

6.16 Photograph by Ronald Patrick

7.01, 7.10, 7.12–7.15 Photographs by Nolberto Salinas González (Noler)

7.02 Roberto Segre, *Diez años de arquitectura en Cuba revolucionaria* (La Habana: Ediciones Unión, 1970)

7.03 *Ciclón*, directed and produced by Santiago Álvarez, 1963. Instituto Cubano del Arte y la Industria Cinematográfica

7.04–7.06, 7.08 Photographs by Patricio Núñez

7.07, 7.09 *Auca* 23 (1972)

7.11 "Mujeres al volante de una grúa," *Paloma* 11 (1972)

7.16 Photograph by Hugo Palmarola Rodríguez

8.01 Martín Noel, Comisión Estética Edilicia, *Proyecto orgánico para la urbanización del municipio. El plano regulador y de reforma de la Capital Federal. Buenos Aires* (Buenos Aires: Peuser, 1925)

8.02 V. Martin de Moussy, *Description géographique et statistique de la Confédération Argentine*, 3 vols. (Paris: Firmin Didot frères, fils et cie., 1860–1873). Courtesy Instituto de Arte Americano e Investigaciones Estéticas "Mario J. Buschiazzo," Facultad de Arquitectura, Diseño y Urbanismo, Universidad de Buenos Aires

8.03, 8.09–8.12 Courtesy Archivo Museo Mitre, Buenos Aires

8.04 V. Martin de Moussy, "Map of the Provinces of Entre-Rios, Santa-Fé, and Banda Oriental," 1865, in *Description géographique et statistique de la Confédération Argentine*, 3 vols. (Paris: Firmin Didot frères, fils et cie., 1860–1873), adapted by Patricio del Real

8.05–8.06, 8.13 Courtesy Archivo General de La Nación, República Argentina

8.07–8.08 Courtesy Centro de Documentación e Investigación de la Arquitectura Pública, Ministerio de Economía y Finanzas, República Argentina

8.14 *Edifici Pubblici e Privati. Opere dell'Ingre. Francesco Tamburini Architetto, Direttore del Dipartimento Nazionale di Architettura della Repubblica Argentina.* Photographs by Samuel Boote, ca. 1886

8.15–8.16 *Vistas de Escuelas Comunes* (Consejo Nacional de Educación, 1889). Photographs by Samuel Boote. Instituto de Arte Americano e Investigaciones Estéticas "Mario J. Buschiazzo," Facultad de Arquitectura, Diseño y Urbanismo, Universidad de Buenos Aires

9.01 Photograph by Mário Fontenelle, courtesy of the Public Archive of the Federal District, Brazil

9.02 *Habitat* 34 (September 1956), courtesy of Museu de Arte de São Paulo, Brazil

9.03 Courtesy of the Museu do Índio, Rio de Janeiro

9.04 Photograph by Heinz Forthmann, courtesy of the Museu do Índio, Rio de Janeiro, Brazil

9.05 Courtesy of the Public Archive of the Federal District and the Subsecretaria do Patrimônio Histórico, Artístico e Cultural, Distrito Federal, Brazil

9.06–9.07 Library of the Faculdade de Arquitetura e Urbanismo, Universidade de São Paulo, Brazil

9.08 Courtesy of Museu de Arte de São Paulo, Brazil

9.09, 9.12, 9.14–9.15 Courtesy of National Institute of Colonization and Agrarian Reform (INCRA), Brazil

9.10 *Manchete*, 1973

9.11 Paulo Tavares

9.13 Courtesy of National Institute of Colonization and Agrarian Reform (INCRA), Brazil; GoogleEarth satellite image, courtesy Paulo Tavares

9.16 Courtesy of United Nations Environment Program

9.17 Photograph by Murilo Santos, courtesy of Instituto Sócioambiental, São Paulo, Brazil

10.01 *Architectural Record* (November 1930), Architectural Record © 1930, The McGraw-Hill Companies

10.02 Courtesy California State Archives

10.03 *Architectural Record* (October 1920): 262, Architectural Record © 1920, The McGraw-Hill Companies

10.04 *The Mark of Zorro*, directed by Fred Niblo and produced by Douglas Fairbanks, 1920

10.05 *Revista de revistas* (October 1932). Courtesy of *Periódico Excélsior*, S.A. de C.V.

10.06 *Architectural Record* (August 1930), Architectural Record © 1930, The McGraw-Hill Companies

10.07 *Cemento* 30 (1929): 4. Courtesy of Archivo de la Academia de San Carlos, Facultad de Arquitectura, UNAM

10.08 *Cemento* 26 (1928). Courtesy of Archivo de la Academia de San Carlos, Facultad de Arquitectura, UNAM

10.09–10.10 *Forma* 4 (1927): 44, 43. Courtesy of Antonio Ruiz Archives

10.11 *Architectural Record* (September 1931): 162, Architectural Record © 1931, The McGraw-Hill Companies

10.12–10.14 Courtesy The Architectural Book Publishing Company

10.15 *Cemento* 36 (1930). Courtesy of Archivo de la Academia de San Carlos, Facultad de Arquitectura, UNAM

11.01, 11.04 *Proa* 22 (April 1949)

11.02, 11.06–11.09 Jorge Arango and Carlos Martínez, *La arquitectura en Colombia. Arquitectura colonial 1538–1810. Arquitectura contemporánea: 1946–1951* (Bogotá: Editorial Proa, 1951)

11.03 William Ospina and Sady González, *Bogotá 40 años* (Bogotá: Revista Numero Ediciones, 1999)

11.05 Photograph by Andrés Téllez T.

11.10 *Arquitectura y Construcción* 16 (September 1949)

11.11–11.14 *Arquitectura y Construcción* 4 (March 1946)

11.15 *Arquitectura y Construcción* 11 (December 1947)

12.01–12.02 *Arquitectura/México* 13 (July 1943): 181, 184

12.03 *Arquitectura/México* 20 (April 1946): 302–303

12.04 *Arquitectura/México* 37 (March 1952): 127

12.05 *Arquitectura/México* 64 (December 1958)

12.06 *Arquitectura/México* 58 (June 1957): 98–99

12.07 *Arquitectura/México* 94–95 (June–September 1966)

12.08 *Arquitectura/México* 72 (December 1962)

12.09 *Arquitectura/México* 93 (March 1966): 45

12.10 *Arquitectura/México* 60 (December 1957): 198

13.01–13.14 Félix Candela architectural records and papers, Avery Architectural and Fine Arts Library, Columbia University

INDEX